# ALIEN ENCOUNTERS:

# *- The Truth -*

By H. J. Risch & R. Risch

**ALIEN ENCOUNTERS UP TO 2022**

**Alien Encounters – The Truth – by Herbert Risch & Richard Risch**

**This is a non-fictional work based on incidents, field investigation, physical research, interviews, verifiable sources and their information, and various types of physical evidence regarding the phenomena of Unidentified Flying Objects, Unidentified Submerged Objects, and other related topics.**

**Copyright © 2022**

**All rights reserved, including the right to reproduce this book or portions thereof in any form.**

# Table of Contents

**Forward** .................................................................................................................... 10

**Chapter I: Basic Laws of Physics Pertaining to Stars, Planets and the Forming of Life** ........ 18

    Fact: Life Among the Stars ................................................................................... 18

    The Drake Equation: ............................................................................................ 20

    Another Exercise in Mathematics – Types of Stars ................................................ 22

    Goldilocks or Life Zone ........................................................................................ 22

    Fact: Red Rain of India ........................................................................................ 24

    Fact: (SETI) Search for Extraterrestrial Intelligence .............................................. 25

        Problem 1: Propagation Loss of Electromagnetic Waves ................................. 26

        Problem 2: Speed of the Electromagnetic Waves ........................................... 27

        Problem 3: Modulation Techniques ................................................................ 28

    SETI's "WOW" Signal: ........................................................................................ 30

        Problem 1: The "WOW" signal and Hydrogen's Resonant Frequency: ............ 31

        Problem 2: The "WOW" signal and the Radio Astronomy Band: .................... 31

        Problem 3: Bad Equipment Design: ................................................................ 31

**Chapter II: Human Development – The Alien Connection** ............................................. 33

    Rumor - The Battle of Dulce: ............................................................................... 33

    Fact: The United States Army is NOW Training for Subterranean Warfare ........... 37

    Fact: Basic Laws of Physical Development pertaining to Intelligence for Interplanetary Travel ........................................................................................................................ 38

    Fact: Anthropology - Development of Humans ..................................................... 39

        Evolution of Humans Chart ............................................................................. 42

        Homo Sapiens Age (Sapiens meaning wise or Intelligent) ............................... 43

        Middle Palaeolithie Period or Stone Age – 400,000 to 250,000 ...................... 43

        Upper Palaeolithie Period ............................................................................... 43

        SUB-CHART .................................................................................................. 43

    Neanderthal Mutation verses Genetic Anthropology ............................................ 47

    Types of Extraterrestrial ....................................................................................... 48

    Fact: Unexplained Early History of Ancient Mankind ........................................... 52

        Theology: Adam & Eve .................................................................................. 52

Rumor: Atlantis the Lost Continent ......................................................................... 53

Fact: Australia ........................................................................................................ 55

Fact: Nasca Plains ................................................................................................. 56

Fact: Lake Titicaca ................................................................................................. 56

Rumor: The Sphinx ................................................................................................ 57

Speculation: About the Sphinx .............................................................................. 60

Fact: The Pyramids ................................................................................................ 61

Fact: The Inca Engraved Stones ........................................................................... 61

Fact: The Rock ...................................................................................................... 62

Fact: The Star Child ............................................................................................... 63

Fact: The Maps ...................................................................................................... 65

## Chapter III: The Modern Era of UFOs Sightings ..................................................... 68

The First Recorded Accounts ................................................................................ 68

Fact: The Battle for Los Angeles ........................................................................... 69

1943 - 1944; During World War II .......................................................................... 72

NASA Mission STS 51A ........................................................................................ 76

June, 1947; Captain Smith ..................................................................................... 77

June 21, 1947; Maury Island, Puget Sound, Washington State ............................ 77

On June 24, 1947; Kenneth Arnold ........................................................................ 82

July 3 – 4, 1947; an object crashed in White Sands, New Mexico known as the Roswell Crash 83

Fact: January 7, 1948; The First Pilot Lost in Pursuit of a UFO ............................ 84

July 24, 1948; Eastern Airlines Flight 576 ............................................................. 95

SWORN TESTIMONY - KOREAN WAR ................................................................ 95

    Air Incident: September, 1950; -Source: NICAP/Blue Book - Pilot: Unnamed. ........................ 95

    Ground Incident: May, 1951, near Chorwon, North Korea (60 miles north of Seoul) ................ 97

July 9, 1952; Two Pan American Pilots ................................................................. 98

July 19, 1952; Washington D.C. (July 19, to July 20, 1952): ................................. 98

May 18, 1953; A UFO Crashed .............................................................................. 101

Rumor: Presidential Visitation ................................................................................ 101

Fact: 1955; Landing at Edwards A.F.B. – Gordon Cooper ..................................... 101

1955; Captain William Coleman ............................................................................ 102

On November 3- 4, 1957; in Levelland, Texas .................................................................. 102

May 20, 1957; in Monstown, England .................................................................................. 102

On December 6, 1958; Over the Gulf of Mexico ................................................................. 102

On April 24, 1964; in Sornoro, New Mexico ........................................................................ 103

Rumor: Landing at Holloman A.F.B. ..................................................................................... 103

On August 3, 1965; Outside of Santa Ana California ........................................................ 105

March 5, 1966; Minot Air Force Base ................................................................................... 108

March 5, 1966; Lieutenant Colonel Salas of SAC ............................................................... 108

March 6, 1967; Security at Francis E Warren Air Force Base in Wyoming ................... 108

September, 1967; Chase over S-2, Offiut Air Force Base ................................................. 109

Fact: October, 1967; Shag Harbor, Nova Scotia ................................................................. 110

On March 24, 1969; in Papua New Guinea ......................................................................... 112

On July 20,1969; Apollo 11 Astronaut Buzz Aldrin ........................................................... 112

February 5, 1971; Apollo 14 ................................................................................................... 113

Fact: Near Collision over Ohio .............................................................................................. 113

In May of 1974; Fort Hancock, New Jersey ........................................................................ 114

Fact: January 8, 1978; Alien Shot and Killed at McGuire Air Force Base .................... 116

Fact: December 24 to 26, 1980; Rendlesham Forest / Bentwaters Sighting ............... 117

Fact: The Official Lieutenant-Colonel Halt Report ........................................................... 124

Fact: Cash – Landrum Incident; December 29, 1980 ....................................................... 126

Fact: Released Documents from a 2010 AATIP Report .................................................... 129

Rumor: 1982; Baltic Sea Area, Lake Baikal ........................................................................ 130

Fact: JAL Flight 1628; November 17, 1987 ......................................................................... 130

September 15, 1991; STS-48 .................................................................................................. 133

September, 1994; Zambagwa, Africa .................................................................................... 134

March 14, 1997; Phoenix, Arizona (The Phoenix Lights) ................................................. 134

2001; Tactical Air Command, Langley Air Force Base ..................................................... 136

2004; Catalina Island ............................................................................................................. 137

November 7, 2006; O'Hara International Airport at Chicago, Illinois ......................... 138

July 2007, Varginha, Brazil: .................................................................................................. 139

Rumor: November 20, 2013 .................................................................................................. 139

July 9, 2015; USS Trepang ..................................................................................................... 139

FOX 32 News, Chicago: Published May 30, 2016 .......................................................... 140

May 19, 2017; Manatouba, Canada .......................................................................... 141

July 15-16, 2019; Off of Coast of San Diego, California ............................................ 141

Secret United States Air Force Vehicle; September 22, 2021 ................................... 144

## Chapter IV: Roswell – The Cover up ........................................................................ 147

Fact: Crash at Roswell ................................................................................................. 147

Fact: Air Force's Initial Reaction to the Roswell Crash .............................................. 148

The fog of Project Mogul: A litany of secrets and misinformation ........................... 151

Fact: Radar Screens ..................................................................................................... 155

    Interviewee: Frank Kaufmann - Army Counter Intelligence Corp Agent ............ 155

Fact: 2nd Wreckage Site – Major Crash Site .............................................................. 156

Rumor: Number of Bodies .......................................................................................... 157

Fact: Civil Engineer (2nd Hand Information) ............................................................. 157

    Interviewee: Vern and Jean Mattais ...................................................................... 157

Fact: The 1st Wreckage Site – Minor Crash Site ........................................................ 158

    Interviewee: Bill Brazel (Son) ................................................................................. 159

Fact: The 1st Wreckage Site - Continued ................................................................... 160

    Interviewee: Major Jessie Marcel ........................................................................... 160

    Interviewee: Doctor Jessie Marcel Jr. (son) .......................................................... 162

Fact: The Press Release to the Public ......................................................................... 162

    Interviewee: Lieutenant Walter Haut ..................................................................... 162

"RAAF Captures Flying Saucer on Ranch in Roswell Region" ................................. 163

Fact: Civilian News Report .......................................................................................... 165

    Interviewee: Lydia Sleppy ....................................................................................... 165

Fact: W.E. Whitmore, Owner of KGFL Radio ............................................................. 166

Fact: Chaves County Sheriff's Office ......................................................................... 167

    Interviewee: Phyllis McQuire .................................................................................. 167

Fact: Fire Chief's Daughter ......................................................................................... 168

    Interviewee: Frankie Rowe ...................................................................................... 168

Fact: The Roswell Mortician ....................................................................................... 169

    Interviewee: Glenn Dennis ..................................................................................... 169

Fact: The Pilots who flew the wreckage out of Roswell .................................................. 171
    Interviewee: Sappho Henderson .................................................................................. 171
    Interviewee: Robert Sherkey ........................................................................................ 171

Rumor: Presidential Visitation ............................................................................................... 171

Rumor: Other Presidential Visitations .................................................................................. 172

Fact: Secretary of Secrets ..................................................................................................... 175
    Interviewee: Charles Welhielm – Handyman – 2nd Hand Information ........................... 175

Fact: Roswell Crash 2nd Site Continued ............................................................................... 176

Fact: Where some of the crash material went ..................................................................... 179

Fact: The Material Found at the Crash Site ......................................................................... 182

A blend of Fact and Unsubstantiated Account: The Material Found at the Crash Site - Continued .............................................................................................................................. 186

Fact: Later Investigation ........................................................................................................ 189

Unsubstantiated Account: Alien Autopsy .............................................................................. 190

**A Roswell Staff Officer Claims There Was A Cover-Up** .................................................. 196

# Chapter V: Alien Abductions, .............................................................................................. 198

# Teleportation VS Quantum Physics ................................................................................... 198

Rumor: Attempted Teleportation ........................................................................................... 198

Fact: Quantum Teleportation ................................................................................................. 198

Fact: The Abduction of Betty and Barney Hill – Case # 1 ................................................... 203
    Interviewee: Betty Hill and Marjorie Fish ..................................................................... 203

Fact: Small Aliens Reported in Newspaper Column in 1955 .............................................. 207

Fact: Travis Walton – Case # 2 .............................................................................................. 208

Fact: The Pascagoula Abduction – Case # 3 ....................................................................... 214

# Chapter VI: The Government and Related Material ....................................................... 218

Rumor: The Aurora Project – Dark Star ................................................................................ 218

Rumor: The TR3B Black Manta - "The Bat" .......................................................................... 219

Fact: The Strange Death of James Forrestal, Secretary of Defense ................................. 221

Rumor: The Black Pyramid .................................................................................................... 226
    Investigating Agencies of the Government on UFOs .................................................. 227
    Types of UFOs Sighted ................................................................................................... 227
    Some Criteria Noted ....................................................................................................... 230

Fact? Bob Lazar .................................................................................................................. 231

Fact: The Black Knight ...................................................................................................... 234

    Early History of the Black Knight ................................................................................. 235

    The First Photographs of the Black Knight ................................................................... 237

Rumor: The Moon Monolith .............................................................................................. 239

Fact: The Martian Monolith ............................................................................................... 239

Fact: Astronauts coming forward about UFOs .................................................................. 240

Multiple Confirmations of Alien Technology ...................................................................... 243

Fact: The Piney Woods Incident, Cash-Landrum .............................................................. 243

Rumor: "Solar Warden" ..................................................................................................... 245

Fact: Sonic Weapons for Defense & Mass Beachings ...................................................... 247

Fact: Strange Electromagnetic Fires and Other Things .................................................... 250

Fact: HAARP (A weapon for Defense against UFOs?) ..................................................... 253

FACT: The Ionosphere ...................................................................................................... 256

FACT: Vibrations, Piezoelectricity, Earthquakes, and HAARP ......................................... 258

    Now we come to earthquakes…and HAARP. ............................................................. 259

FACT: Invisibility & Cloaking Now Possible ...................................................................... 262

## Chapter VII: Beyond Earth - Mars .................................................................................. 263

The Fascination of the Red Planet .................................................................................... 263

The Lost Martian Space Probes ........................................................................................ 264

The Case of the Mars Bio-Station ..................................................................................... 266

The "Face" of Cydonia ....................................................................................................... 274

Quoted from NASA Report: ............................................................................................... 277

The Pyramids of Cydonia .................................................................................................. 277

The Other Faces of Mars ................................................................................................... 278

Martian Photographic Mysteries Taken by Curiosity ......................................................... 278

Another Mystery of Mars ................................................................................................... 288

A TRUE ANALYSIS OF THE PHOTOGRAPH OF THE MYTERIOUS LIGHT .................. 292

## Chapter VIII: Beyond Earth - The Sun vs Venus .......................................................... 296

Things Far Beyond Technology & Imagination ................................................................. 296

## Chapter IX: Misdirection, Hoaxes, Insufficient Data, Misidentifications and Misconception 305

Fact: The Twining Memo – Misdirection ...... 305

Fact: Gorman's Battle – Insufficient Data ...... 306

Fact: George Adamski - Hoax ...... 307

Fact: The Condon Committee – Misdirection ...... 308

Fact: Billy Meyers - Hoax ...... 309

Fact: Gulf Breeze – Hoax? ...... 310

Fact: Cattle Mutilations - Misidentification ...... 311

1945 December, Florida: Flight 19 - Misconception,...yet a Mystery ...... 313

FACT: Judge Graham Stikelether (Main Figure in Story) ...... 314

The Story: ...... 315

The Story goes on: ...... 316

1963 - The Continuation of the Flight 19 Story: ...... 318

The Finding of the .50 Caliber Machinegun: ...... 318

The Teaming up of Graham Stikelether and Tom Myhre ...... 321

Fact: Getting Serious about Sirius - a Misconception ...... 329

## Chapter X: Notable Movies and Television Specials Recommended for Viewing ...... 332

Fact: UFOs Are Real ...... 332

Fact: Documentary Television Special: UFO Cover-Up Live ...... 333

Fact: Alien Secrets - Area 51 ...... 335

Fact: UFO Files: Alien Engineering ...... 338

Fact: I know what I Saw ...... 340

Fact: Out of the Blue ...... 342

## Chapter XI: An Analysis And Still More Questions ...... 344

Analysis ...... 344

More Questions ...... 347

# Forward

In regards to UFOs (Unidentified Flying Objects), agreeably there have been numerous hoaxes, mistaken identifications, and unreliable if not total "kooks" seeking their 15 minutes worth of fame by concocting outlandish stories of UFO encounters. However, the overwhelming majority of witnesses; which encompass police officers, pilots, astronauts, military personnel as well as observant average citizens; can hardly be lumped into this former group. Added to this are the thousands of hours of video recordings taken at live events with either cell phones or various other types of professional recording equipment (*such as United States Navy fighters' gun cameras*). This is what this book is about, the quest of discerning true facts from rumors, misinformation, and pure fiction.

One has to then ask a simple question: what is the motivation behind the words of skeptics who are outwardly and so ardently opposed to any type of exploration, and scientific analysis, or any comprehensive research into UFOs and Aliens? Are they more concerned with the loss of their own reputations because the facts undercut any theories or opinions they may have espoused previously? Are they a part of an organized disinformation program that seeks to obscure the facts of truth from the general public, or are they merely individuals with lack of foresight out of fear causing them to live in denial that there are more intelligent beings in the universe then us?

These individuals appear to lack any capacity for common sense or any leaps of rational logic and thought. One also has to ask, *what credentials do they really have that classifies them as knowledgeable individuals or experts on this subject? How open are they to overwhelming physical evidence that indicates UFOs are truly real? And how can they be so certain of their opinions and explanations if they never observed what was reported...or at least taken the time to go to the location where the incident occurred and search for clues and evidence along with collecting witnessed accounts? Seems as though these cynics have an agenda,...one not based on truths.*

Our manuscript was written with a motive too. It differs from any other book on this subject matter because it was specifically designed and penned to demystify all the irrational expounding of nonsense belched out by these skeptics. For these hostile so-called '***debunkers***' have never been taken to task over their poor excuses and lame explanations that they elucidate. By any intellectual titles they may hold, these brash cynics have bullied good law-abiding people into silence and for the sake of "mudding up the waters" of honest testimony to stymie further investigation. However, by well-established scientific technical methods and principles, common sense, and plain hard facts, we will endeavor to expose the truth by which these debunkers have attempted to conceal so obviously.

Through time-tested techniques, plausible if not verifiable scientific explanations and facts, we will dissect the most astounding controversial and bizarre cases on record. Also, in asking the most provocative and challenging questions to cut away the veil of deception, these simply queries might sway any doubting opinions...*to new possibilities*.

We believe no other book has ever done this before,...but perhaps it will start a trend. For when the light of truth is shined upon deceit, a lie cannot hide.

However, do keep in mind, although both co-authors held security clearances with the United States military and government (*and both having our own separate UFO encounters*), we are only part-time researchers. Most of these compiled mysteries, compelling tales, and associated data have already been presented to the public by devoted and accredited investigators on this subject over time.

The material contained within is chronologically formatted so that a clear and concise picture and understanding made available to the reader. We have also touched upon the sciences of Acoustics, Anthropology, Astronomy, Astrophysics, Electronics, Etheric and Dynaspheric Forces, Geology, Geophysics, History, Mathematics, Magnetism, Medical Procedures, Meteorology, Mythology, Paleontology, Quantum Physics and Theology in an effort to supply reasons for our conclusions. Facts will be presented along with their source of origin whenever possible. Rumors on the other hand are only the retelling of what is unsubstantiated, it should be noted that in many rumors, nearly always contain the seeds of truth...*if not far more than that*.

Speculations and tended conclusions are those of the most likely of possibilities, but cannot be counted as the only ones or for that matter taken as proof. Each will be clearly identified. So let us begin our journey in seeking the truth about UFOs.

Three final noteworthy news events we have listed, however, before proceeding any further in this book. The first occurred in 2008. The Catholic Church *surprisingly recognized the existence of extraterrestrial biological beings* as per the Vatican astronomer Father José Gabriel Funes. This is a huge stride for the religion that once condemned Galileo's observations about the solar system.

The second, a more recent and far more interesting noteworthy news item, was reported January 6, 2014 by Fidel Martinez from an interview Canada's former defense minister, Paul Hellyer, publicly stated for the first time *that aliens are quite real!*

He stated, "I've been getting from various sources (that) there are about 80 different species which consist of 4 categories and some of them look just like us and they could walk down the street and you wouldn't know if you walked past one."

*What if this were all true, eighty different races with a good handful of them **human in appearance**? Surely the former minister of Canadian defense would know having worked hand-in-hand with the United States government. I wonder what the skeptics have to say about this revelation coming from a man once entrusted with his nation's deepest and darkest military secrets. But maybe these skeptics consider him one of those kooks too. Just like J Edgar Hoover of the FBI, who requested by way of an official 1947 FBI memo (acquired through the Freedom of Information Act) access to the flying saucer that crashed in Roswell, New Mexico. **(See image below.)** By the way, this request was made partially in his own handwriting, so its authenticity is guaranteed. By the way, the memo is dated within two weeks **AFTER** the crash at Roswell!*

Memorandum for Mr. Ladd

Mr. ▓▓▓▓▓ also discussed this matter with Colonel L. R. Forney of MID. Colonel Forney indicated that it was his attitude that inasmuch as it has been established that the flying disks are not the result of any Army or Navy experiments, the matter is of interest to the FBI. He stated that he was of the opinion that the Bureau, if at all possible, should accede to General Schulgen's request.

SWR:AJB

### ADDENDUM:

I would recommend that we advise the Army that the Bureau does not believe it should go into these investigations, it being noted that a great bulk of those alleged discs reported found have been pranks. It is not believed that the Bureau would accomplish anything by going into these investigations.

DML

(Clyde Tolson)

I think we should do this
7-15

(J. Edgar Hoover)

I would do it but before agreeing to it we must insist upon full access to discs recovered. For instance in the Sw. case the Army grabbed it & would not let us have it for cursory examination.

H

However, the third and most compelling admission that UFOs are indeed real was announced in 2007 by the United States Navy, confirming the authenticity of three F-18 gun-camera videos taken in 2004 of a flying "*Tick-Tack,*" which could both outmaneuver and outrun the Navy's fighters.

Similarly, another incident confirmed by ship borne radar captures and videos in 2019 was first released by The New York Times in 2021 as well as by the Mystery Wire web site and many UFO research organizations and other media outlets showing separate incidents of "Unidentified Aerial Phenomena," or UAPs (*a more correct term than UFOs*). One can clearly see the objects maneuvering and appear under intelligent control on the film released. There was also the admission that the Pentagon had operated a secret UFO investigatory project, called the *Advanced Aerospace Threat Identification Program* (AATIP) over many years.

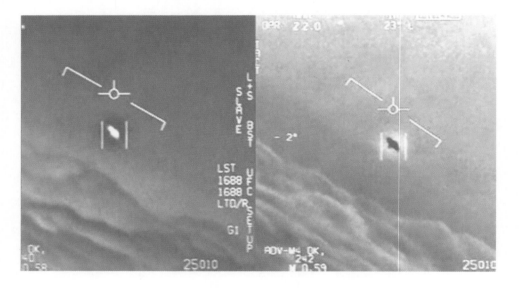

To be clear, the Navy is *not* saying that these videos show evidence of alien life. Rather, the Navy is saying it can't identify the phenomena in the photographs and videos. However, to be certain, these were not the only "Unidentified Aerial Phenomena" ever captured on by other military equipment. Incredible footage from a night vision camera below shows an encounter captured by a seaman aboard a destroyer, the *USS Russell*, in its deployment with the 7th Fleet in the Philippines Sea.

**Question:** *Who's considered <u>not a kook</u> by these skeptics and so-called experts, <u>only those who don't disagree with them</u>?*

**Answer:** Such a determination would suggest a bias by these critics that the United States Navy and the Pentagon are...*a bunch of kooks*, wouldn't it? It would seem so when ridicule, denigration, and absurd answers replaces reasoning in any attempted at adult and intelligent dialogue or conversation seeking to seriously examine the UFO phenomena from a truly scientific *vantage point*.

Dismissing something as *"swamp gas," "a flight of birds"* or *"the planet Venus"* gets old real fast especially when credible witnesses and hard evidence are accompanied by in-depth critical analysis that proves to the otherwise...*or are these skeptics and so-call "experts" calling the Pentagon and the United States Navy liars for providing solid hard evidence that UFOs exist?*

*And these are not the first admission either made by the United States Military that UFOs are real.* On September 23, 1947, General Nathan Twining, head of the *U.S. Air Material Command* (AMC) wrote a classified letter to Air Force General George Schulgen. The letter explicitly says that UFOs are very much real and an unexplained phenomenon. This memo was uncovered by the late UFO researcher and Nuclear Physicist Stanton Friedman "in a classified box in a classified vault."

```
Memo C. Form No. 10-414 (Rev 10 Sep 46)        NND 760168  5-4-73           WF-LJ JAN 47  2008
                                          CAS   SECRET

                                    HEADQUARTERS
IN REPLY ADDRESS BOTH               AIR MATERIEL COMMAND
COMMUNICATION AND EN-
VELOPE TO COMMANDING                                                TSDIN/HMM/ig/6-4100
GENERAL, AIR MATERIEL                                               WRIGHT FIELD, DAYTON, OHIO
COMMAND, ATTENTION
FOLLOWING OFFICE SYMBOL
                                                                    SEP 23 1947
  TSDIN

         SUBJECT:   AMC Opinion Concerning "Flying Discs"

         TO:        Commanding General
                    Army Air Forces
                    Washington 25, D. C.
                    ATTENTION: Brig. General George Schulgen
                               AC/AS-2

            1. As requested by AC/AS-2 there is presented below the considered
         opinion of this Command concerning the so-called "Flying Discs". This
         opinion is based on interrogation report data furnished by AC/AS-2 and
         preliminary studies by personnel of T-2 and Aircraft Laboratory, Engineer-
         ing Division T-3. This opinion was arrived at in a conference between
         personnel from the Air Institute of Technology, Intelligence T-2, Office,
         Chief of Engineering Division, and the Aircraft, Power Plant and Propeller
         Laboratories of Engineering Division T-3.

            2. It is the opinion that:

                 a. The phenomenon reported is something real and not visionary
         or fictitious.

                 b. There are objects probably approximating the shape of a
         disc, of such appreciable size as to appear to be as large as man-made
         aircraft.
```

And the United States is also not the only government to do so either. On September 17, 2016, over the Bristol Channel, a British police helicopter belonging to the National Police Air Service (NPAS) in South Wales photographed the object below, a glowing orb, using an infrared camera set to pick up heat-emitting objects. The police have openly admitted to taking the video of the object posted below, which they have no explanation for.

**Question:** *Are these skeptics and so-call "experts" now calling the British Police Force of South Wales a bunch of kooks and/or liars for their providing of solid hard evidence that UFOs exist?*

**Question:** Does this make the point that *these skeptics and so-call "experts" are the real highly unreliable kooks and liars trying to dismiss hard evidence with irrational excuses or any other insane pretext that crosses their minds?*

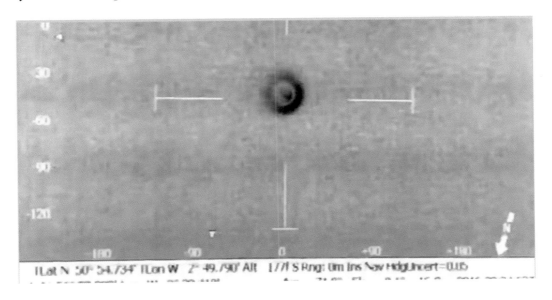

Going back to Mr. Hellyer, if his claims do indeed prove to be the truth, it would explain a lot. In addition, Mr. Hellyer went further…by listing where these aliens are coming from. It turns out that these alien visitors have been popping in from as near as a base on one of Saturn's moons, the far side of our moon and star systems such as the Pleiades and Zeta Reticuli (*which just coincidently are mentioned by other witnesses in this book describing their own UFO encounters…and ordeals*). He also admitted that the aliens are sharing their technology through some sort of arranged agreements with the Canadian and United States governments. How fascinating that he should validate the very stuff that conspiracy theorists have disclosed. We do intend to explore every aspect of this too…and much more. And to this end, we cite a forgotten quote:

> *Truth will ultimately prevail where there is pains to bring it to light.*
>
> — George Washington —
> Commander-in-Chief of the Continental Army and 1st President of the United States of America

# Chapter I: Basic Laws of Physics Pertaining to Stars, Planets and the Forming of Life

Since July 7th, 1947; there has been speculation on the existence of alien races visiting our planet. In this non-fictional book and report, we will attempt to analyze all the information which has been accumulated over the years regarding the logical criteria for extraterrestrial visitations. This will include types of extraterrestrial, abduction, types of craft, known visitations and possible reasons for visitations.

We will also debunk those theories which defy logical interpretation and reasoning. Some of the material covered will be drawn from our first documentation of several years prior (*Rumors of UFOs, fact or fiction*) since that research is still pertinent, factual and grounded in reality. Repeating this previous statement, the material contained within this book is chronologically formatted so that a clear and concise picture and understanding has been made available to the reader.

## *Fact: Life Among the Stars*

In starting our honest search for UFOs, it begins with a simple question:

**Question:** *What is the possibility of other advanced intelligent life existing somewhere else in the universe?*

Every astrophysicist agrees that the Milky Way, our galaxy, is approximately 100,000–120,000 light-years in diameter. The age of the Milky Way is thought to be 12 to 16 billion years old. Our star, the Sun, may be found close to the inner rim of Orion's Arm, towards the end of one of the Milky Way's spirals.

Our planet, the Earth, is found in the habitable zone of our own solar system. In astronomy and astrobiology, the habitable zone is the region around a star where it is theoretically possible for a planet with sufficient atmospheric pressure and temperature to maintain liquid water upon its surface. However, this presupposes that all life requires water ($H_2O$) to exist.

With these facts in mind, consider this; in 2003, a shortlist was created by astrophysics students at the University of Arizona, which consisted of 30 stars from an estimated 5,000 stars *within* 100 light years of Earth suitable for hosting complex life forms (and civilizations). Part of the criteria used in this academic exercise was that the life forms had to be *at least equal* to our own civilization in terms of technology. As you can imagine, finding those 30 planets just within 100 light years of Earth was astounding!

Using these numbers from a point of view of percentage alone (and most certainly *not using* the Drake Equation), the calculated amount of 0.6 of a one percent (30/5000) seems insignificant at a glance. However, inasmuch as there are somewhere between 100 billion to 400 billion stars in the our galaxy, the Milky Way, the number of possible places where life is equal to or in advance of our own is not small. The estimated number from using the above calculation is from 600 million to 2.4 billion planets hosting individual alien civilizations capable of space travel! Whether it is just traveling to an orbiting moon or traveling to other stars, those numbers it is assumed are sprinkled generously with both—and everything in between. With an estimated 10 billion stars having planets orbiting in the habitable zone of their parent stars, it makes perfect sense. Some of these planets are considerably older than the Earth, while others are of approximately the same age or slightly younger.

**Question:** *Isn't it possible we are being visited by aliens from at least one of these 30 neighboring planets mentioned above?*

**Answer:** I would give this a resounding "*YES!*" Of these 30 stars, two particular candidates come up in several notable UFO cases discussed later in this book. These base stars are named Zeta 1 and 2 Reticuli. Located in the skies of the Southern Hemisphere of our planet, they can be only view by telescope from places like Brazil.

**Question:** *Is this just a quirk of fate considering the comments made previously by Mr. Paul Hellyer? What do you think?*

Nevertheless, there are those who insist that there is no life other than our own, or if there is life, it is not capable of higher technology. This is purely arrogance. No matter how you figure it mathematically, you will always come up with an answer that testifies *we are not alone*...and high probability that they more than just a little technically advanced then us. It makes you wonder if some of these so-called skeptics suffer from the defects of a inferior public school educational system.

## *The Drake Equation:*

The Drake Equation is an equation used to estimate the number of detectable extraterrestrial civilizations in the Milky Way galaxy. The equation was devised by Frank Drake: an Emeritus Professor of Astronomy and Astrophysics at the University of California.

The Drake Equation states:

N (Number of Civilizations Capable of Interstellar Communication) = R (Rate of Star Formations) * FP (Fraction of Stars with Planets) * NE (Number of Planets suitable for Life) * FJ (Fraction of Planets where life can develop) * FI (Fraction of Planets with Intelligent Life) * FT (Fraction of Planets with Intelligent Life that are capable of interstellar communication) * L (Lifetime of Intelligent Civilization that is capable of interstellar communication)

**NOTE:** *Only the R factor is known in this equation and only included the Milky Way Galaxy. This only gave a number of 10,000 of Communicating Civilizations.*

This formula was updated by Amir Aczel to include all galaxies in the universe which are described in his book "Probability ONE"

Below is a new formula that is used in some Earth - Space Science Manuals:

NPL (Number of Planets with Life) = NC (Number of Chances) x NS (Number of Stars) x FP (Fraction of Planets) x NLZ (Number in Life Zone) x FL (Formula for Life) x FA (Formula for Atmosphere) x FC (Formula for Correct Conditions) x UF (Unknown Factors)

$$NPL = NC \times NS \times FP \times NLZ \times FL \times FA \times FC \times UF$$
or the derived amount of civilizations as
$$1{,}260{,}000 = 4 \times 1\text{ Trillion} \times .35 \times \underline{.06} \times 1 \times 1 \times 0.75 \times 0.00002$$

Not a small amount, is it?

The Drake Equation is closely related to—and vexed—by the Fermi Paradox. The Fermi paradox suggests that even though a large number of extraterrestrial civilizations would form, however; these same technological civilizations would tend to disappear rather quickly and should act as a Great Filter to reduce the final value of the Drake Equation.

According to this scenario, either it is very hard for intelligent life to arise, or the lifetime of such civilizations must be relatively short. This again is only a premise: an exercise in thought. Part of the Fermi Paradox requires these aliens to think and behave (and to be as self-destructive) as human beings under similar environmental conditions.

**Question:** *If aliens evolved under different environmental conditions of their worlds, would they tend NOT to think and act like human beings?*

**Answer:** Of course and therefore they would come up with different solutions (governed by the capabilities of their technologies and the adhesion of their societies) to find workable solutions to address any catastrophic disasters. Human beings have the traits of self-interest and invent out of necessity. However, this can be a self-destructive trait when ignoring others in our society in indulging in our own egos. On the other hand, a "hive" mentality and genetics may be the key to the longevity of a species survival, but not without the tradeoff of diminished creativity and a rigid society with a predominance towards conflicts with other species to the point of annihilation of one or the other.

To be honest, there will be a certain amount of alien civilizations that will disappear due to self-induced or natural disasters. (Perhaps…one day our own will.) There is no question of that. But then again, we—the authors—tend to lean more towards that number being far, far fewer than the Fermi Paradox dictates. And if you want proof of this, all you have to do is look at mankind's own history.

**Question:** *How many times has the human race rebuilt itself from catastrophic disasters?*

**Answer:** Many times! Life is tenacious and resilient. We as a species have out-survived dinosaurs, meters slamming into the Earth, ice ages, plagues, floods, droughts, and even our own arrogance and stupidity that brings about war. *And if the aliens retain the same basic instinct to survive that all other living things possess, therefore, there should be a much greater number of alien civilizations that have endured and multiplied in the aftermath of such calamities.* But keep in mind, this was all in the past—when we didn't have the capacity for self-annihilation or dealing with the numerous sociopaths and psychopaths in power. *The combination of both has the potential to lead to a very violent and disastrous end.*

As you can plainly see the Fermi Paradox is so blind-sided, it should be rendered to history as a footnote of intellectual nonsense—and something to be joked about. (Give it another hundred years and it will be—*if it's remembered at all*.)

Going back to the Drake Equation, it stated that given the uncertainties, **the original equation** estimated that there were probably between 1000 and 100,000,000 civilizations in the galaxy. We think the latter number—which mirrors our example of using percentages mentioned above—to be the more accurate one, which may be proven out by the additional mathematics illustrated below.

## *Another Exercise in Mathematics – Types of Stars*

Stars are classified by their spectra (the elements that they absorb) and their temperatures. There are seven main types of stars. These are listed in the order of decreasing temperatures. They are: O, B, A, F, G, K, and M.

## *Goldilocks or Life Zone*

The Goldilocks or habitable life zone is that distance from the base star where temperatures on a planet's surface would not be extremely hot or extremely cold for liquid water to exist. This would naturally depend upon the type of base star the planet is orbiting.

Types of stars to be discarded from the equation of life are as follows:
 4 % = Blue Dwarfs
13 % = White Dwarfs, Red Dwarfs, Black Holes and Neutron Stars

43.4 % = Red Giants

32 % = K Stars (with the exception of some warmer burning K stars - see below)

92.4 % is the number of star systems in the universe that are unable to support life.

This leaves only 7.6 % (which is the total number of all G stars) of all star systems in the universe having the capability to support life. Multiple this by the number of stars in the universe, estimated to be one trillion and combine the other factors in the Drake Equation you arrive at 1,260,000 stars that have the capability of planets able to support life.

As stated above, G type stars, *like our Sun*, make up about 7.6% of all stars. This is a relatively small amount considering the percentages of the others. However, the average G star longevity is 90% greater than most of the other stars and has a luminosity class of V. Such a star has about 0.8 to 1.2 solar masses and a surface temperature of between 5,300 and 6,000 Kelvin. Like other main-sequence stars, a G - V type star is in the process of converting hydrogen to helium in its core by means of nuclear fusion. Our Sun is the best known (and most visible) example of a G - V type star. Each second, it fuses approximately 600 million tons of hydrogen to helium, converting about 4 million tons of matter to energy. Other G - V type stars include Alpha Centauri A, Tau Ceti, and 51 Pegasi. However, some warmer K stars sequence range 0 to 1 fall within the temperature zone that permits the existence of $H_2O$ (water) which is necessary to produce life. Cooler F stars that sequence range falls between 8 to 10 may also have the possibility to produce water and thus develop life.

Let's approach the amount of planets harboring life in our galaxy in another way using a hypothetical assumption that life can only evolve on planets orbiting G type stars. (Of course life can evolve on planets orbiting other types of stars, but we are using this purely for the sake of argument.) Since there are somewhere between 100 billion to 400 billion stars in our galaxy, therefore the amount of planets with evolving life forms orbiting around G type stars (counting every one of them of course) should be between 7.6 billion (*or 7.6 % of 100 billion*) to 30.4 billion (*or 7.6 % of 400 billion*). And even if we were to take *the calculated amount of 0.6 of a one percent* (remember that magical number derived from the study of neighboring stars done by the University of Arizona), it would still come out to be 45.6 million (*or 7.6 billion x 0.6*) to 182.4 million (*or 30.4 billion x 0.6*) alien civilizations equal to or in advanced to us in terms of technology. These numbers are strangely in agreement with (*probably between 1000 and 100,000,000 civilizations*) the high-end number in the *original* Drake Equation. Another coincidence mayhaps?

See the relationship between the highest number from the Drake Equation and our simple mathematics? The Drake Equation number sits directly in the center of our derived calculations, and our numbers were simply the resultant from the amount of G-type stars in the galaxy combined with 0.6% from the study by the University of Arizona. From this it is easy to come to the conclusion that the deduced minimal amount of existing alien technological civilizations right now *is at least* 45.6 million.

It doesn't take an abstract formula to prove a hypothesis where simplified mathematics will do just fine.

**On March 6, 2009 the Kepler telescope was launched to study that section of the constellation known as Orion's Arm. Its main purpose was to determine if there were any planets in that section of that star system). To date, more than 4,000 exo-planets have been discovered and are considered "confirmed" (and counting).**

## *Fact: Red Rain of India*

In the city of Kerala, India, from the 25th of July to 23rd of September, 2001, a torrential monsoon downpour of *red* rain soaked the region. This strange phenomenon was first recorded in Kerala a few hours after an airburst by a meteor, which exploded in the atmosphere above the outskirts of the city. More than 120 such rain showers were reported in that year, including yellow, green, and black ones.

Investigated by Astrobiologist Godfrey Louis, pro vice-chancellor at nearby Cochin University of Science and Technology (CUSAT), the samples of red rainwater were studied and discovered to contain strange properties, including auto fluorescence—light that is naturally emitted by cell structures like mitochondria. However, what was most astounding was that these red rain samples from 2001 contain a thick suspension of cells, which although lacked DNA, could still replicate itself as Godfrey Louis observed. It was further noted by Louis, *"Unlike other biological cells, these red rain microbes can withstand very high temperatures. It is possible to culture them at temperatures as high as 300 degrees centigrade (572 degrees Fahrenheit)."*

Louis, believing that these cells could be only extra-terrestrial in origin because of existing theories, which had already hypothesized that comets may have a hot water core with chemical nutrients able to support microbial growth, immediately contacted NASA. However, NASA refused to even talk about it and dismissed any notion of further investigation into the matter without comment.

Astrobiologist Godfrey Louis and his colleagues at Cochin University of Science and Technology are now investigating just how dangerous these cells are to human beings and to our bio-systems.

**Question:** *Why would NASA **not be concerned** about possibly a deadly alien microorganism that could potential kill off all life on the planet?*

**Answer:** Maybe they already know what it is.

Accordingly via several televised broadcasted television shows, two intelligence agents (*supposedly from Area 51 and written about later in this book*) disclose an agreement and exchange program between the United States government and an alien race. This agreement has been purportedly in existence for the last twenty years with these aliens from the base star Zeta 1 Retucli. If true, maybe part of this exchange is biological knowledge.

## *Fact: (SETI) Search for Extraterrestrial Intelligence*

*<u>In our humble opinion</u>, unless it has some other hidden purpose,* one of the biggest wastes of time, effort, and money has got to be the **search for extraterrestrial intelligence** program, otherwise known as SETI. This dubious scheme (<u>*in our humble opinion again*</u>) was either conceived as some sort elaborate smoke screen by the US government for the purpose of disinformation on UFOs, or a way to obtain finances to

pay for black projects...other than printing up a few trillion (*worthless*) *fiat* dollars by the federal reserve (*a privately owned central not really a part of the government*) bank.

Anyway, SETI is run by Harvard University, the University of California, Berkeley and the SETI institute. It is a wonder why such esteem universities would ever consider signing on to the SETI Institute and what we consider a seemingly ridiculous approach for alien contact. This perceived comedy of errors can be easily seen in the principles involved with the program.

SETI's goal is to monitor various frequency ranges, and has done so over many years, with the intent to intercept non-human "intelligent" broadcast signals. Before being *defunded* by the United States government in the 1980s, the range extended from 1200 MHz to 3000 MHz (Mega Hertz). This is well within the UHF band of frequencies that can be used for television broadcast, and considered part of the microwave region of the electromagnetic spectrum. Some of these frequencies are also assigned specifically for radio astronomy usage only. Overall, these frequencies were chosen for the "quietness" of the range from electromagnetic noise generated by the rest of the universe. However, herein lay several problems as to why this is so unusable...and so ludicrous in our view.

**Problem 1: Propagation Loss of Electromagnetic Waves**

Electromagnetic waves are a set of **combined and associated** electric and magnetic fields, in which each field is always at right angles to the other as it propagates (travels) through air or space. They cannot be separated until they are received by broadcasting equipment whose antennas they are directed at. To do so would destroy them...and any intelligent information they may be carrying.

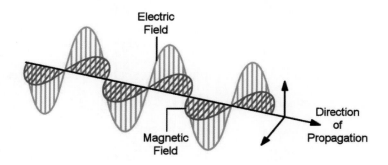

As an example, say we need to broadcast an audio signal of a change of orders to one of our starships 10 light years away. First this requires imposing a voice signal (the range of which would contain human voice frequencies between 20 hertz to 30 kilohertz) upon a higher carrier frequency. Let's make that 1420 MHz. This electronic combining is performed through two processes called *mixing* and *modulating*. After the modulation signal is sent through an amplifier, it is then passes to an antenna array to be broadcasted into space.

The first problem encountered is called *propagation loss*. Regardless of how powerful this communication signal is, the further our spaceship is away from the Earth transmitter source, *the more the signal tends to disperse*. If you want to test this theory out for yourself, put on any AM radio station and drive away from the transmitter.

Say the transmitter is located in New York City and you're driving down Route 80 going west to Pennsylvania. The further you get from NYC, the more the signal strength decreases. Volume of the broadcast signal tends to diminish, becoming eventually distorted, next breaking up, and final finally dissolving into static or be overwhelmed by another and closer AM radio station on the same or any nearby frequencies (which is *jamming* what left of the original signal).

This illustrates perfectly what would happen to our signal broadcasted to our starship. The further it moves away from our planet, the more it diminishes into the background noise of the universe, until it cannot be distinguished from the background noise itself. And unless they flunked physics, the SETI scientists should also know this as well.

**Question:** *If SETI scientists know that any intelligent radio signal will disperse so dramatically that it can't be picked up, why even bother in listening for one?*

**Answer:** Maybe they have an ulterior motive. We simply don't know. To ignore the basic principles of electronic broadcast raises more than just an eyebrow.

**Problem 2: Speed of the Electromagnetic Waves**

Electromagnetic waves travel slightly slower than the speed of light. Again, this is pure physics. So let's think about that. We have a starship that can create a wormhole

(*Rosenbridge per Albert Einstein*) and perhaps travel nearly instantaneously to any point in space. It is currently sitting 10 light years away from the Earth and ready to receive a change in orders from us. However, it would take 10 light years for that signal from the Earth to get it to them via electromagnetic waves and then 10 more years for an acknowledgement reply. In other words, a total of 20 years wasted in sending *just one* communication signal!

**Question:** *How inefficient is that?*

**Question:** *Do you think any starfaring race would use such antiquated technology for communications to distance starships or outposts? Let's be realistic about this.*

In David Weber's *Honor Harrington* science fiction series, he addresses this problem. For distant communications, David Weber uses small courier boats to shuttle communications back-and-forth vast distances while using electromagnetic broadcast signals for *localized solar system communications only*. This would probably be the first actual method used in the human colonization of space. *Doesn't this make perfect sense?*

In the science fiction thrillers *Beyond Mars Crimson Fleet* and the upcoming book *Beyond Mars Dark Matter*, the Martian Fleet relies upon *communication buoys* much like today's cell towers as repeater stations. The difference being that these buoys *exist partially in dimensional hyperspace itself*, transmitting and converting tachyon particles that are thought to travel theoretically much faster than the speed of light along with having a dimensional time component as well. (*It may be possible to receive a tachyon communication signal before it is even sent!*) Still this is mere conjecture, but it does emphasize a point. These two communication options demonstrate advance technology that would be absolutely required by any starfaring race, and not the absent broadcast signals expected by the SETI Institute. As we have stated, we truly believe SETI is a complete waste of time, effort, and money.

**Problem 3: Modulation Techniques**

There are multitudes of modulation techniques which are available and in use today by the human race. These include, but are not limited to: Amplitude Modulation, Frequency

Modulation, Pulse Amplitude Modulation, Pulse Coded Modulation…the list goes on. However, to receive any signal, you must have the hardware equipment compatible with the method. This being the equipment set at the right frequency with the correct demodulation and decoding scheme to interpret the information contained within the signal. (You cannot receive or decipher an Amplitude Modulated signal if you are using Frequency Modulation equipment as another example, and if it is encoded, the problem increases 10 fold.) If anything is lacking or improperly set, you cannot determine between the actually reception of an intelligent communication signal (if you receive anything at all) or looking at some sort of space phenomena such as a pulsar (which was once thought to be an actual alien communication signal).

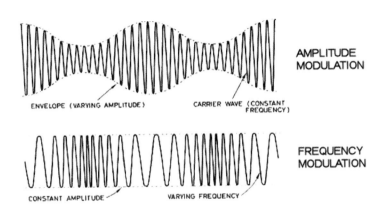

And what if an alien civilization is utilizing a totally different modulation method that we have no knowledge of? What if they're using some multi-frequency and randomly coded modulation scheme that consciously prevents and defeats eavesdropping? You do realize there is at least one group of what *they might consider hostile aliens* in the cosmos that may pose a dire threat to their people—*us!*

**Question:** *Do you still think that SETI could intercept and interpret any signal reliably to determine that another civilization really exists?*

**Answer:** It should be a resounding, "Hell, no!" We, ourselves, believe that SETI's equipment would better be served as part of a warning system looking for planet-killing asteroids rather than looking for non-existent signals that can never be intercepted.

## SETI's "WOW" Signal:

The final argument against SETI has got to be the famous "WOW" signal. The "WOW" signal was a strong narrowband signal detected on August 15th, 1977 which lasted for approximately 72 seconds. This highly questionable *transmission* burst was intercepted by Jerry R. Ehman using the *Big Ear* radio telescope (located at Ohio Wesleyan University's Perkins Observatory). Although this signal did last 72 seconds and had all the earmarks and expected signature of an extraterrestrial signal, it was never again picked up by SETI regardless of numerous attempts they made to do so. There are several obvious reasons as to why nothing came of these attempts.

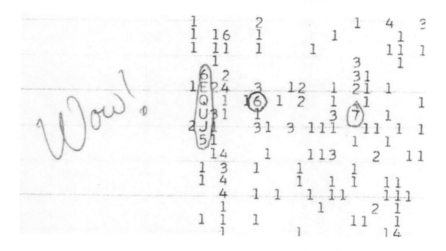

**NOTE:** (June 7, 2017) A team of researchers with the Center of Planetary Science (CPS) finally solved the mystery of the "Wow!" signal from 1977. They reported that it was a comet—one that that was unknown at the time of the signal discovery. Lead researcher Antonio Paris described their theory in the *Journal of the Washington Academy of Sciences*. It was noted that the frequency was transmitted at 1420 MHz,...*which just happens to be the same frequency at which hydrogen vibrates at.*

The explanation—the signal originated from a comet whose trajectory took it away from the earth, and therefore, would explain why the signal was not heard again. The scientist then noted that two comets had been in the same part of the sky that the Big Ear was monitoring on the fateful day. The comets now identified as P/2008 Y2 (Gibbs) and 266/P (Christensen) were only recently discovered.

*Now when you read "Problem 1" below, you'll find that we, the authors of this book, theorized about this occurrence in the publication of our prior book "<u>Rumors of UFOs</u>" on July 30, 2014! <u>That's 3 years earlier than the Center of Planetary Science's conclusion!</u>*

**Problem 1: The "WOW" signal and Hydrogen's Resonant Frequency:**

*The most obvious problem in frequency to the "WOW" signal was that it matches the resonant frequency of hydrogen.* The Upper and Lower Sidebands of the signal (1420.356 MHz and 1420.4556 MHz respectively) are the same distance apart (0.0498 MHz) as the hydrogen element line (using a spectrometer or spectrum analyzer). **Since everything in the universe came from hydrogen particles, it could be assume this burst was the result of a coronal mass ejection (CME) or some other space phenomena regarding hydrogen.**

**Problem 2: The "WOW" signal and the Radio Astronomy Band:**

To the uninformed, 1420 MHz just happens to be within the band of frequencies used by radio telescopes only.

**Question:** *Do you think that there is a possibility that some other observatory might have been calibrating its equipment and transmitted the "WOW" signal accidentally?*

Back of the heyday of Amplitude Modulation, the military used what was called "signal bounce" to extend the range of communication signals. Signals were routinely bounced off of the ionosphere, but this same effect can be achieved by bouncing a signal off of some orbiting space junk as well. This could be a possible explanation as to why such a strong signal was only heard once.

**Problem 3: Bad Equipment Design:**

The most blatant flaw discovered while investigating the "WOW" signal, lies in the design of the *Big Ear* telescope itself. The telescope used two feed horns (microwave antennas), each pointing in a slightly different direction of the sky following the Earth's rotation. This is not unusual, however. What is unusual is the fact that the electronic engineer/s who designed this system neglected to insert any circuits to tap each horn feed individually so as to determine which one was receiving a signal and at what strength. Apparently he or she never heard of a technique called *direction finding*, which defeats the purpose of not only the telescope…but the SETI program itself. (*It was even suggested that the signal picked up was inadvertently the electronic noise made from a nearby rotating beam from a lighthouse. Really?*)

This little design error made it virtually impossible for anyone to ascertain from where the signal originated. Although both were pointed at the constellation of Sagittarius at the time, the sky is still a pretty damn big place (*the closest star of Tau Sagittarii being 122 light years away*).

**Question:** *Are you still wondering why SETI hasn't contacted any aliens yet? Maybe SETI would be better off building a campfire and using a blanket to send smoke signals instead?*

**Skeptics:** There are no aliens, if there where why would they be visiting our world, what would they want here?

**Speculation:** An advanced race of aliens would want to learn about any civilization and if that civilization presented a future threat to them. They may also have the curiosity to investigate other worlds and other civilizations. If their home planet was in jeopardy or perhaps they for some reason cannot reproduce, then they would need either a new world to relocate their race to or another species to develop a hybrid race to ensure survival. Of course there just may well be other motivations (either benevolent or malevolent), which may vary with any alien species we encounter.

**NOTE**: Recently our scientist have discovered indications that four thousand exo-planets with some of them similar to our earth. Exo-planets are planets that orbit a star outside our solar system. They have categorized in four sections. These are Gas Giants, Neptunian. Super-Earth and Terrestrial; the latter refers to those planets most like our own.

# Chapter II: Human Development – The Alien Connection

## *Rumor - The Battle of Dulce:*

Since both authors have serve in elite units of the United States military, by our experience it is hard to judge some events where little or no evidence or information exists. It doesn't necessarily mean that something is made-up, but rather...*craftily hidden*. The battle of Dulce falls into this category. Some details have the ring of truth, while others seemed a little far-fetched.

The supposed 1979 battle was purportedly between elements of United States military's special operation groups against as many as a contingent of 37 alien species, who had broken an agreement with the United States government. Accordingly, the aliens were holding as many as 3500 humans (*mostly female*) for experimentation.

What took place out there (if anything) is a matter of conjecture that has been played up by a few televised broadcasts and other media. This incident might have also been alluded to in an interview with Bob Lazar, former Area 51 scientist, who made mention of a small group of security personnel having a shootout with a group of aliens. But again, he might have been referring to something else within the same time frame.

With that all said, *yes*, there are indications (physical evidence) that there was a base of some type located at Dulce, New Mexico and there are physical (*magnetic and radiation*) anomalies associated with the location. But perhaps this is all that can be discerned since it was said that the overwhelming majority of the base was constructed underground with elaborate tunnels and labyrinths. And one must remember that the United States government had many such places built for different purposes, e.g. for strategic fuel reserves, and NORAD (North American Aerospace Defense Command located under Cheyenne Mountain), which monitors all air traffic in our atmosphere to maintain security and secrecy.

Yet, the most compelling proof that such a battle did take place was a man by the name of Phil Schneider, purportedly one of three survivors from the battle. And what made him most unique was his *extensive* battle scars: the fingers of his left hand that had the appearance of being burned off with high electrical voltage discharge and the huge scar that ran down his chest, **neither none of which can be accounted for by any medical records available.**

**NOTE:** *One of the authors over the years had known three veterans, two of which served in the 101st Airborne Division while the third was a special forces operations veteran. All sported recognizable military ordinance wounds not traceable or supported by any medical records, military, civilian or otherwise.*

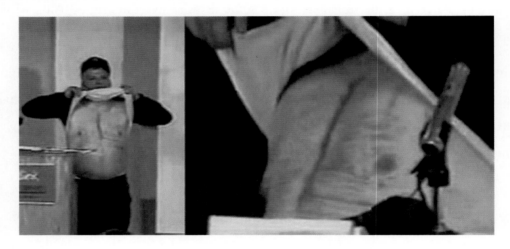

In 1996, Phil Schneider, however, was found dead in his apartment to what appears to be an execution style murder. The man had been strangled by his own breathing tube from a Portable Oxygen Machine wrapped around his neck. And according to sources, it appeared that he repeatedly suffered torture before he was finally killed. Seven months prior to his death, Schneider did lectures on the clandestine work he had discovered at Dulce. Peculiarly, his death was listed as a suicide, regardless of the evidence suggested murder to the contrary.

**NOTE:** *For two years prior to his death, Philip Schneider had been on a lecture tour talking about government cover-ups, black budgets, and UFOs. Philip stated in his lecture that in 1954, under the Eisenhower administration, the federal government decided to circumvent the Constitution and form a treaty with extraterrestrials. The treaty was called the 1954 Greada Treaty. Officials agreed that for extraterrestrial technology, the Grays could test their implanting techniques on select citizens. However,* *the extraterrestrials had to inform the government just who had been abducted and subject to implants. Slowly over time, the aliens altered the bargain, abducting and implanting thousands of people without reporting back to the government...as so is spoken in rumors and whispers.*

**NOTE: The government memorandum below is seen by many ufologists** as evidence that the First Contact meeting had occurred with extraterrestrials with a distinctive 'Nordic' appearance. However, supposedly the extraterrestrials were spurned due to their refusal to enter into technology exchanges while insisting on nuclear disarmament by the US and presumably other major world powers. This in turn started a series of meetings that led to a treaty eventually being signed with a different extraterrestrial race dubbed the "Greys"—*and other extraterrestrials.*

**Office Memorandum** · UNITED STATES GOVERNMENT

TO: Mr. A. H. Belmont
FROM: Mr. C. R. Roach
DATE: September 24, 1957

SUBJECT: UNIDENTIFIED FLYING OBJECT REPORTED ON SEPTEMBER 20, 1957; INTELLIGENCE ADVISORY COMMITTEE - WATCH COMMITTEE

Reference is made to my memorandum of September 23, 1957, on the captioned matter which reflected an Intelligence Advisory Committee (IAC) evaluation of an Air Defense Command (ADC) report that radar stations at Montauk Point, Long Island, New York, and Benton, Pennsylvania, detected an unidentified object proceeding in a westward direction with an altitude of 50,000 feet and a speed of 2,000 knots (approximately 2,300 miles per hour) on September 20, 1957. The IAC intelligence evaluation of this report reflected that "It is highly improbable that a Soviet operation is responsible for the unidentified flying object reports of September 20, 1957."

Today, at an Executive Session of the IAC, General Millard Lewis, Director, Air Force Intelligence, advised that although the ADC has not completed its investigation of the evidence on this matter, there are continuing indications that the object detected was an atmospheric phenomenon. The radar pickups now reflect speed variations in the object's course, ranging from 1,500 miles per hour to 4,500 miles per hour. This latter speed is improbable according to U. S. scientific theory for any type flying object which this could conceivably be. General Lewis added that the present sun spots are associated with the peculiar radar activity throughout the globe and that this could have some cause for the captioned report.

ACTION:

Liaison will report further information on this matter as quickly as it is developed.

MWK:bjt
(7)
1 - Mr. Belmont
1 - Mr. W.C. Sullivan
1 - Mr. D.E. Moore
1 - Mr. Whitson
1 - Liaison Section
1 - Mr. Kuhrtz

*In 1979, Philip was employed by Morrison-Knudsen, Inc. He was involved in building an addition to the deep underground military base at Dulce, New Mexico. The project at that time had drilled four holes in the desert that were to be linked together with tunnels. Philip's job was to go down the holes, check the rock samples, and recommend the explosives to deal with the particular rock. In the process, the workers accidentally opened a large artificial cavern, a secret base for the aliens known as Grays. In the panic that occurred, sixty-seven workers and military personnel were killed, with Philip Schneider being one of only three people to survive. Philip claimed that scars on his chest were caused by his being struck by an alien weapon that would later result in cancer due to the radiation.*

## Fact: The United States Army is NOW Training for Subterranean Warfare.

*Currently, the United States Army is learning to fight underground, spending over $572 Million at this new training initiative. The Army is currently pushing to train soldiers how to fight in subterranean tunnels, mines, sewers, underground complexes and other environments around the globe. The question is why?*

If Philip Schneider's claims were true, then his knowledge of the secret government, UFOs and other information kept from the public, could have serious repercussions to the world as we know it. In his lectures, Philip spoke on such topics as the Space-Defense-Initiative, black helicopters, railroad cars built with shackles to contain political prisoners, the World Trade Center bombing...***starting with the collapsing of Building Seven 30 minutes before the attack actually began***, and the secret black budget.

**Skeptics:** Why would aliens visited our world as they would come in contact with unknown bacteria, fungus and disease?

**Speculation:** Aliens having the ability to travel through or bend space would already be aware of this factor and would have sent out probes to get atmosphere, soil, and plant samples for analysis to determine what affect our world would have on them before visiting our planet. Our own civilization has sent probes to the Moon and Mars to learn about their conditions. Would not an alien race to the same to any world they wish to learn more about. In our past, there have been numerous UFO reports of probes collecting samples of our soil and plant species.

**Skeptics:** If there are aliens from other worlds why haven't they contact the world governments or tried to invade our planet as seen in so many Science Fiction movies?

**Speculation:** Perhaps these aliens have developed beyond our primitive rationality to a more benevolent society seeking to blend cultures and ideas. However, this does not mean we should completely disregard ulterior motives for their presence.

There have been all kinds of extraterrestrials reported by individuals to the authorities. However, the most common species are the Grays and Humanoids that somewhat resemble us.

## *Fact: Basic Laws of Physical Development pertaining to Intelligence for Interplanetary Travel*

In some science fiction movies they show aliens with tentacles similar to squid or octopus. This is some writers copy from H. G. Wells novel "War of the Worlds". This concurs with an obscure legend of a race called "the Nemos", but again, this is just a legend...*for now*.

This on the surface seems to be a total fallacy for the simple reason that in order to create any complex vehicle one would think hands and fingers were needed. With that said, however, we do not know if it such things were/or are possible. Unlike UFO critics, we acknowledge that things might be beyond our understanding and be completely plausible unless shown evidence to the contrary. Nature always seem to have a way.

For the most part, we do believe a being must have thumbs and fingers. These we consider necessary to construct all the minute pieces that these vehicles need. Therefore logic demands that these beings be humanoid.

Another fact is that these beings must be carbon based in structure. Life here on earth has an atomic carbon based structure of #6 = 1/2/1 + 1/4/1 with the ability to link onto itself.

Silicon has an atomic structure closest to our own and has a possibility to link onto itself with the based structure #14 = 1/2/1 + 1/8/1 + 1/4/1, thus, the ability to form life. It is not known if any silicon life forms are in existence. The problem with silicon life forming is that silicon atoms do not bond together easily and are weaker when bonding does occur, thus it is less likely to form complex chains of atoms which is needed to create life.

Basic laws supporting intelligent life apply everywhere in the universe. Scientists claim that intelligent life can evolve in any form even that which we cannot recognize. This may be true, however, in order for a being to travel through the void of space, a being must be encased in a craft that can protect it from the extreme cold, unknown radiations and other dangers that the journey may present. It must also have a contained breathing source since no atmosphere exist in the ether of space. This requires the being to create intricate and minute parts in the construction of its vehicle, which will allow it to make the journey to other worlds.

In order to create these complex devices a being must be able to manipulate appendages to create, assemble and pickup the small intricate materials needed to comprise such a complex design. Thus, any being possessing the intelligence to travel through the vastness of space, must have some sort of fingers and thumbs and would most likely maintain some sort of humanoid form. Their atomic structure would probably be that of carbon based.

## *Fact: Anthropology - Development of Humans*

Africa is considered the birthplace of human lineage. Human evolution is supposedly the product of physical and behavioral morphology. However, Anthropologists have uncovered so many variations of our species in different stages of development during the evolution of mankind, it is still very unclear how they all are pieced together.

**Question:** *Could each variant have been the testing of prototypes by aliens in order to reach our current species of Homo Sapiens?*

The earliest stage of humanoid beginnings has been traced back to the Ramapithecus genera of huminidae. This creature lived about 13,000,000 years ago and was related to both ape and man. This determination was fostered primarily by their dental resemblance (their jaws and teeth being similar to human beings).

The next later genus was labeled Australopithecus huminidae or Near Man who died out about 2,000,000 years ago. These creatures were the size of chimpanzees, but had more of a human form. They were also believed to have more of a potential of using stones as tools (although chimpanzees have been discovered using natural objects as crude tools) which meant that they were developing some deeper form of reasoning. These ancestors walked upright with a bipedal gait and had a brain 1/3 size of present day humans. An anthropologist named the remains of one of these ape ancestors that emerged from the trees in Ethiopia, Lucy (after a popular Beatles' Song).

Other anthropologists then uncovered skeletal remains of Homo Habilis or "*the Handy Man.*" This species of humanoid was found with stone tools and was believed to be first of the Homo genera. They dated back to 2,300,000 years ago. They were followed by Homo Sapiens or True Man, which consisted of several classifications of huminidae.

The first genus was Homo erectus, which was split into two groups: Java Man and Peking Man, who both lived about 1.8 million to 70,000 years ago. The second genus was Homo Heidelberg (*the inventor*) living 800,000 to 300,000 years ago. The third genus was Homo Neanderthal living about 400,000 to 30,000 years ago. It is believed that this species of human branched off from Homo erectus. The fourth genus was Cro-Magnon Man. Cro-Magnon Man was also termed Progressive Neanderthal, since Homo Neanderthal was dying out. This was believed to have occurred around 50,000 to 30,000 years ago and was also believed to be one of the direct ancestors to modern men.

However, there are those anthropologists that argue True Man and Near Man are not truly related and evolved from separate branches. In other words True Man was developing while Near Man was still in existence. But there are further indications that True Man and Near Man were from completely different lineage.

The human brain is twice as large as that of chimpanzees and gorillas, which started with the development of the Homo Habilis. The increase in size allowed social learning to occur with the development of crude language. This is due to a disproportionate enlargement in the temporal lobes and prefrontal cortex during this period of human development. But the cause for such development is still not understood. Although some anthropologists speculate that an increase of meat to their diet may have caused this mutation. But that doesn't make sense, chimpanzees are also omnivores. Why would similar species with similar diets evolved so differently? *Therefore, something else had to have happened, something more compelling to produce such a radical affect.*

**NOTE:** An interesting genetic trait that few human incur is **Hypertrichosis**. Also called **Ambras syndrome**, it is an abnormal amount of hair which grows over the entire body. Extensive cases of hypertrichosis, informally called **werewolf syndrome**, are extremely rare (approximately 100 cases reported world-wide in 1995. This gives a logical link to our primal ape past, but how much of a link? With approximate + 7 billion people on our planet, shouldn't there be more than just 100 cases? It means that something happen to greatly alter our genetic code where the growing of bodily hair was reduced considerably, which strangely coincided with enlargement of the human brain.

To add to this puzzle, researchers at the University of Chicago looked at variations of genes regarding modern human brains such as microcephalin and ASPM respectively. They found evidence that the two genes have continued to evolve. For each gene, one class of variants has arisen recently and has been spreading rapidly. For microcephalin, the new variant class emerged about 37,000 years ago and now shows up in about 70 percent of present-day humans. For ASPM, the new variant class arose about 5,800 years ago and now shows up in approximately 30 percent of today's humans. These time windows *are extraordinarily short in evolutionary terms*, indicating that the new variants were subject to very intense selection pressure that sped up their evolution in a very brief period of time…both well after the emergence of modern humans about 200,000 years ago. However, the most starting surprise to all this is that there is no definitive explanation as to why no other animal species on the face of the Earth has had any similar development. Could it be that someone (of Extraterrestrial origin) deliberately manipulated human DNA?

And it gets better. There are a host of underwater cities that exist such as off of the island of Yonaguni, near Okinawa, Japan; off the shores of Alexandria, Egypt; and off the shores of the city of Dwarka, India that are said to be anywhere from 4,000 to 32,000 years old! And there are more of them; at least 200 sites in the Mediterranean Sea alone have been identified. These are not counting another 200 that lay on the various **ocean floors**. This means human civilization is considerably older than once thought…and the evolution of humans considerably faster than imagined!

**Question:** *By Darwin's logic, how can that be? How is this all possible? If diet is the only explanation, why didn't the Gigantopithecus (giant ape of the Himalayas and China that existed from 13 million to 500,000 years ago weighing approximately 600 to 1,000 pounds) intelligence increase because they were also omnivorous?*


**Evolution of Humans Chart**

**Development Years        Classification        Further Information**

65 Million Years Ago – Plesiadapis – Mammal similar in appearance to a Ferret

13 Million Years Ago – Ramapithecus – 1st hominid much like a Gibbon in appearance as their teeth and jaw bone were similar to Human and Gorilla.

8 Million Years Ago – Gorillas linkage breaks off from Human linkage

7.2 Million Years Ago – Schelanthropus – Partial Bipedal

5.7 Million Years Ago – Orrorin Tugenensis

5.6 Million Years Ago – Ardipitheans – Full bipedal

4 Million Years Ago – Chimpanzee linkage breaks off from Human linkage

4 Million too 2 Million Years Ago – Australopithecus – (Lucy found in Ethiopia)
* Brain size begins to increase
* Length of pregnancy extends as does child maturity
* Fertilization of female becomes year round with no outward signs (swelling of vagina as in chimpanzees) that female is fertile.

2.3 to 1.4 Million Years Ago – Homo Habilis (Handy Man)
* Brain becomes 2 times larger than Ardipitheans
* Beginning of tool making stage
* Earliest fossil of atomically correct human develops. (Homo – means humanus given by Carolus Linuaeus as method of classification).

## Homo Sapiens Age (Sapiens meaning wise or Intelligent)

1.8 Million to 70,000 Years Ago – Homo Erectus – Examples: Java and Peking Man,
* Brain capacity doubled – 1st to use fire and 1st to leave Africa.

800,000 to 300,000 Years Ago – Heidelberg Man – Invented needles, slicing and striking devices.

## Middle Palaeolithie Period or Stone Age – 400,000 to 250,000

400,000 to 30,000 – Neatherthal Man – Interbred with other human ancestors – 4% of Neatherthal genes were found in non-African present day humans.

## Upper Palaeolithie Period

50,000 to 40,000 Years Ago – Cro-Magnon Man
* Used flint tools, knives, blades, fishing hooks
* Had rituals and ceremonies such as burying their dead
* Created jewelry and cave drawings and wore clothing (*because of loss of hair*)
* Bartered among small groups, wore clothing
10,000 Years Ago – Modern Man – Rising of civilizations and agriculture.


## SUB-CHART

| | | |
|---|---|---|
| Plesiadapis | | 65,000,000 years ago |
| Ramapithecus genera | | 13,000,000 years ago |
| Sahelanthropus - died off at | | 5,700,000 years ago |
| Orrorin Tugenensis - died off at | | 5,600,000 years ago |
| Ardipitheans - died off at | | 4,000,000 years ago |
| Australopithecus genera or Near Man – died off at | | 2,000,000 years ago |
| Homo Habilis – developed at 2,300,000 | to | 1,400,000 years ago |
| Homo Erectus – developed at 1,800,000 | to | **70,000 years ago** |

(Now at this point a 20,000 year gap in Evolution appear in between the genera. No skeletal remains whatsoever have been uncovered to show this progression.)

| | | | |
|---|---|---|---|
| **Heidelberg Man – developed at 800,000** | Sub | to | **300,000 years ago** |
| **Neanderthal Man - developed at 400,000** | Species | to | **30,000 years ago** |
| **Cro-Magnon Man - developed at 50,000** | | to | **40,000 years ago** |

( Another 30,000 year gap in Evolution appear in between the genera. No skeletal remains uncovered to show progression.)

Modern Man – developed *supposedly* at **10,000 years**    to    Present Day

The timeline development of Modern Man (once assumed to have arisen 10,000 years ago) is now in question. There are still new discoveries being made in our linage every year. As a result, the data mentioned above may not be up to date.

**NOTE:** The Legend of Atlantis, the Lost Continent, was said to exist during this time period. This has been given credibility by the 200 sites of underwater cities found around the world's oceans that are believed to be dated at least 32,000 years old. And they are believed to have been submerged as ice caps melted at the end of the last ice age approximately 9-10,000 years ago. Dwarika (or Dwarka Gujarat) was the name of this supposedly 32,000-year old underwater city founded by Hindu Lord Krishna, which existed from 32,000 to 9,000 BC. It was rediscovered in the year 2000 (and 120 feet underwater) when a water pollution study was performed near the site. This means a civilization as big as New York's Manhattan Island was in existence prior to the assumption of 10,000 years of modern civilization.

Carbon-14 dating on debris recovered from the site, including construction material, pottery, sections of walls, beads, sculpture and human bones and teeth, put it at nearly 9,500 years old, making it older than the Sumerian civilization by several thousand years. There are also several Hindu texts that clearly state that mankind and his civilization is over 100,000 years old...*and possibly much older*. They also report the usage of what can be described as modern-day weapons…*including the deployment of a nuclear bomb?* What can be believed about this?

Skeptics dismissed this as the product of pure fantasy. But why would the ancient's record fiction when they were more concerned about preserving knowledge and history in their writings? However, to add credence to the Hindu texts, several ancient cities thousands of years old (as an example, one located in Rajasthan, India) have been found to contain unusually high concentration levels of radiation with portions of them showing exposure to intense heat (*in the form of melted rock*). And the one in Rajasthan, India, the radiation is still so intense, the area is considered highly dangerous! Built around 8,000 to 12,000 years ago, the levels of radiation there have registered so high on investigators' gauges that the Indian government has now cordoned off the region. Now what do you suppose could have done that?

There is something else too; the aforementioned over 420 submerged cities have been discovered all around the globe. For example, the mystery of Dwarika is joined by the Yonaguni Monument, a rectangular, stacked pyramid-like monument is believed to be more than 10,000 years old. Sitting off the coast of Japan's southern Ryukyu Islands, it has been nicknamed "Japan's Atlantis." Now how do you suppose it got there?

One has to consider that Neanderthal man and Heidelberg man were a sub species of human and integrated with Cro-Magnon man as was explained earlier in regards to gene analysis and by reviewing the chart above Cro-Magnon man developed while Neanderthal man was still in existence.

**Question:** *So who built these cities?*

**Question:** *Because they are submerged, do you think that would indicate that current sea level was raised over 100 feet from the level it had been previously (possible due to some gigantic flood that was spoken of and recorded in the Holy Bible, Hindu text and other cultures' writings around the world—which lends creditability to these writings accuracy while contradicting the ranting of the skeptic "experts")?*

**Question:** *Wouldn't a gigantic flood around the entire world explain why all previously technology and any historical evidence were wiped out and forced mankind to start all over again maybe around 10,000 years ago?*

**Question***: Another oddity, biblical figures who accordingly lived to well-seasoned ages of 900 years or more, are accompanied by other ancient texts from many cultures, which also have listed life spans of 1,000 years. If true,...what changed that?*

**Question:** *It is possible for the seafloor beneath each of these 420 cities to have sunk or dropped, but...if that were so, why then are the majority of these cities intact? Surely such an event drop the floor unevenly showing cracks and ridges. Such a catastrophe would have made a complete shambles out of all the structures of these cities. If so, then why didn't it?*

**NOTE:** The period of 10,000 year B.C. has another major significance, the bringing of great rain storms upon the Earth with the heating up of the planet. Perhaps the waters of a great flood came from melting ice? This also corresponds to the dated rain water erosion of the Sphinx. We believe the last time it rained on this monument was between 9,000 - 8,000 B.C. (This is discussed a little later on in more detail.)

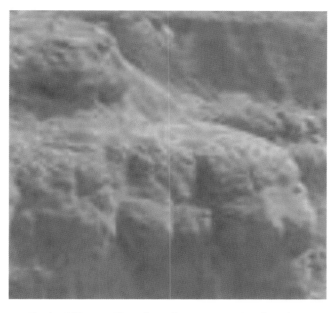

**Rain Water Erosion found on the Sphinx**

## *Neanderthal Mutation verses Genetic Anthropology*

Mutation is said to be one of the causes of evolution along with environmental factors. By consensus, it is believed that hominoids (ancient humans) separated from their ape ancestors because of mutation. This occurrence was purportedly due to a gene that caused a smaller jaw to develop in humans thus allowing the skull cavity to enlarge growing rounder and giving room for the brain to increase. This eventually led to the development of language and tool making. One gene believed to be responsible for this mutation is MHY16, which is predominate in apes giving them thicker and larger jaw muscles. However, human mutation prevents MHY 16 from becoming dominant. Homo Habilis was the first humanoid that showed this cranial and jaw difference, which began the development of a larger brain capacity eventually leading to developing tools. Yet, it would take a million years before the final results of Modern Man to appear.

Mutations can be caused by chemical imbalances within the body and may be a result of damaged DNA, which may not be able to repair itself, causing errors in cellular replication or deletion of that DNA. Mutation can also alter genes preventing them from functioning or function in a different manner causing chromosomes to break-apart or fuse together creating chromosome rearrangement. However, DNA repair mechanisms have the ability to repair most mutated genes and chromosomes before the mutation becomes permanent. It is believed that many of spontaneous mutation is due to DNA replication errors and are considered to be neutral in having no detrimental effect.

Some recent discoveries state that Modern Man is a direct decent from a mutated Neanderthal Man. This theory is known as the Neanderthal Genome Project and is based on analysis of 4 billion pairs of DNA. It states that some mixtures of genes occurred between Neanderthal Man and modern human ancestors of European decent. It also showed that 94 percent of Chimpanzee genes matched human genes not 98 percent as previously believed. However, the DNA range from Neanderthal Man in Modern Man of European decent is 1 to 4 percent and the pattern of mutation sequence differs also there is no evidence of Neanderthal genes in humans of African descent yet their physiology is the same as the European humans. Thus although Neanderthal Man may have interbreed with Homo Erectus it is highly unlikely that modern man is a mutation of Neanderthal Man.

**Question:** *If Cro-Magnon Man developed at 50,000 years ago and is not derived from Neanderthal man or Heidelberg man and Homo Erectus died off at 70,000 years ago then where did Cro-Magnon man arise from?*

**Question:** *If Cro-Magnon Man disappeared at 40,000 years ago and Modern Man arrived 10,000 years ago, then where did Modern Man arise from?*

## *Types of Extraterrestrial*

According to Lieutenant Colonel Wendelle Stevens (United States Air Force Retired) confirms Canada's former defense minister Paul Hellyer assessment that there are several Classification of aliens visiting this planet. He also states these are divided into four categories,

The first are Human or Human Hybreds species which resemble the people of our planet and could pass for earth humans if walking down a street. They are said to want to aid our species.

**Scandinavian Type**     **Cross Breed Type**

The second category are the Grays both small and tall which are from separate solar systems. These seem interested in our genetics and are believed to be creating the human hybrids. Their appearance matches the aliens from the Roswell crash and the abduction of Betty and Barney Hill as well as those reported by Travis Walton in his abduction. They are a pale grey or tan with large heads and almond shaped eyes and slender body framed. They are usually small in height, but there are some reports of very tall Grays most likely from different solar systems. They appear to be vegetarians some emanating very harsh odor when deceased or from any internal body openings from wounds.

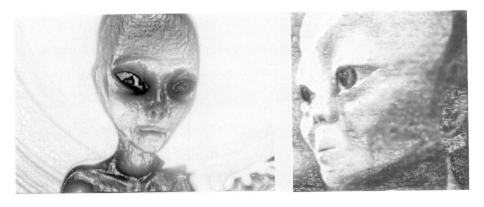

The frame shown directly above was taken seemingly from 8mm **_military film footage_**, which *can neither be proven nor disproven. It remains a controversial mystery to this day.* (This film footage was shot *approximately around 60 to 70 years ago*, at which time this would be very hard to fake,...*more like...nearly impossible from what was viewed.*) This footage represents contact with a lone survivor of a UFO crash.

**Speculation:** The Greys aliens from Zeta 1 Reticuli are reported to be 4 feet in height and have large black retina. In this researcher estimation these attributes may be attributed to their planet's size and distance from their sun. If the planet is larger than our earth it would have a greater gravitational pull thus causing the beings to be smaller in size but with stronger muscles as has been reported in the Hill abduction. If their planet was further from their sun then they would need to have larger retina to see since their world may not have as much light as ours.

The Anunnaki (Deities that were worship by the Sumerians, Akkadians and Babylonians), according to Sumerian text came to our planet from a planet named Nibiru looking for gold around 6,000 years ago. They needed a slave work force to gather the gold so they altered human DNA. This would have put them here during the development of Modern man. It is said that they introduced the Sumerians, to agriculture, writing and mathematics. **They were said to be tall with oblong heads.** They may have been one of the Grays or tan aliens that some sightings have mentioned.

**Ancient Bust of Queen Nefertiti**

**NOTE:** The technical name for this type of human being is, *Homo Sapiens Capensis*.

The third category is said to be Lizard type and are humanoid in body type meaning they are bipedal, but their skin is more lizard-like. They appear to also be interested in our genetics and dispel a temporary paralyzing breath, which they seemed to use when mating with our females.

**Photo Credit: sculpture, Canadian Museum of Nature**

The first three categories has been reported to bring their own nourishments and medical enzymes with them from their worlds. It also has been reported that they communicate through telepathy.

The fourth category is what Lt. Col. Stevens refers to as a catch all, which means that this is a category that doesn't match any of the three main categories. These include, but are not limited to, cyborgs, insect types as seen below and other creatures or life forms that do not bear any resemblance to humanoid life forms.

# Past References

## *Fact: Unexplained Early History of Ancient Mankind*

**Theology: Adam & Eve**

**Speculation:** *Suppose an outside factor was introduced into the equation and interfered with the progress of our species. Perhaps there was never a missing link.*

**Question:** *Is it possible that the apithecus family tree was replaced by or integrated from a form of previous developed humanoids? (See separation above.)*

A segment of the bible tells us of Adam and Eve being the first humans to walk this planet.

**Question:** *Is this biblical reference lore or fact? If it is fact then where did they come from or were they a rapid mutation from Homo Erectus to Modern Man?*

Possible, but highly improbable most mutations would occur spermatic over a lengthy period in time *and not consecutively rapid.*

**NOTE:** The time difference in changes of hominids to humans occurred over a period of at least 100,000 years until the arrival of Cro-Magnon Man and Modern Man where changes occurred in 10,000 years...a rather impossible feat!

**Question:** *The bible states that Adam and Eve were expelled from paradise for disobeying God. If so, where was paradise?*

**NOTE:** As mentioned in the Holy Bible, the Garden of Eden was located where four major rivers converged, including the Euphrates. From that description, Bible scholars conclude that the Garden of Eden was located somewhere in the Middle East area known today as the Tigris-Euphrates River Valley.

**Question:** *Could paradise really have been on another planet in a different star system?*

**Speculation:** Suppose the answer to the above question is correct, and this planet orbiting in another star system had cured its problems of disease and hunger. Everyone on that world had equal wealth and status. Then wouldn't being ejected from this utopia be...*paradise lost*,...like Adam and Eve?

"Ancient Alien Astronauts" documentary from the History Channel speculated that aliens came to earth to mine precious metals and interrupted the evolution of humans to produce a slave labor race. They further believe that our genetic makeup was altered by these aliens to incorporated language and music. Their proof is in all the other animals that developed on this planet around the same time as humans. The proof being that these other animals remained basically in their original form, unchanged and never progressing any major increase in intelligence.

## *Rumor: Atlantis the Lost Continent*

Atlantis was a name given to a supposed continent that had a flourishing civilization, which was said to exist in the 10 millennium B.C. or 10,000 years ago, before the birth of Christ, which coincidently was during the development period of Modern Man (*along with the great flood spoken in previous text*). It was said to be a conquering nation the size of Asia Minor and Libya combined together. Although there are no written manuscripts (*or at least none that we know of*) of this society, their existence was passed down by word of mouth from Egyptian priest to Egyptian priest and finally to Solon an Athenian statesman.

He told the Greek philosopher Plato, who made written dialogues of this civilization in his penned scripts of *Timaeus* and *Critias*. Its exact location was not identified only that it was located in the Atlantic Ocean (*known as the Western Ocean then*) beyond the Pillars of Hercules *(known as the Rock of Gibraltar today)*. It was said to have been destroyed by an earthquake or volcanic eruption (*conflicting stories*) from one to three days, also conflicting information. The Atlantians were said to be a seafaring people having highly evolved technology with not only sea craft, but rumors of air ships that could reach the heavens.

**Speculation:** Let us now presume that on another star system perhaps 5,000 years older than our own, a planet exists which contains an intelligent civilization where all the social problems were resolved. The dilemma of space travel may is a memory of a conquered challenge in their past. Surely these people are still evolving from primitive beginnings.

Yet, this race of benevolent people had to contend with primitive psychological and behavior culminating from their own ancestry consisting of greed, lust and envy. The ruling government being "humane," needed to rid themselves of these malcontents and social outcasts. However, they did not wish to dispose of these anti-social individuals in a harmful way or to lock them away forever.

They then would have chosen another method of disposal. Perhaps they would have then found another world where they could deposit these refugees, but with no method of returning to their home planet.

**Question:** *What if that alien world used our earth in the same manner as that remote island of Australia were used by the British Empire…as a penal colony?*

**Question:** *The problem is that if they were societal outcasts, why? What if these alien social outcasts also had criminal minds and the genetic seeds of sociopaths and psychopaths? What would be the result of a combination of their genes with our ancient genetic code?*

This planet would then be a depository for another world's unwanted individuals. They would have been given the essentials and comforts of survival, but deprived of the technology and materials to return to their home planet. Our planet may have well have been a penal colony they named—*Atlantis*.

No one knows when the demise of Atlantis occurred, however, being a seafaring people it would be reasonable to speculate that some of these vessels and crews would have escaped the catastrophe that befell their country by being at sea or exploring other lands. These survivors would have no other recourse but to interbreed with our primitive race. Therefore, their offspring may have well been the source of not only of inspiring engineers developing the methodical structures of ancient times such as the Sphinx and the pyramids, but also in achieving great power...*to became dictators enslaving the masses*. Our history is filled with that. After all, all things can be a double edge sword.

## *Fact: Australia*

During the 1780's the British Commonwealth had an equally similar situation. Poverty became prevalent and crimes were on the rise. Prisons became bulging with criminals of different magnitudes. Many were of lesser crimes, which only contained individuals for a short period of time before he or she were released back into society. However, after serving their sentence in prison and once more turned loose among the population, those that repeated crime, went to the next level of incarceration.

The solution was the deportation of these individuals to a location in the Pacific Ocean where they could not do any harm to the society of the British Commonwealth. Here, in the remote island of Australia, prison camps were established. The first prisoners were comprised of 520 men and 200 women. They were assigned as laborers working on tracks of land. By 1830 a total of 75,000 prisoners had been banished to the New South Wales Section of Australia for internment in these labor camps.

However, certain food supplies became scarce and Great Britain was forced to provide provisions for their existence (which could not be grown or procured from this island sanctuary). This penal colony was eventually abolished in 1867. Once these criminals had served their penalty they were released, but were forbidden to return to England or its Commonwealth. However, after the completion of their sentence they could apply for and receive land grants.

## *Fact: Nasca Plains*

In Peru, South America there is a section of the Andes mountain range where a plateau exist that is 50 miles long between the cities of Nazca and Palpa south of Lima. The peaks of other mountains still stand, but the peak of this one particular plateau is flat as if it had been shaved off, yet it is reported that there is no rubble on either side of this plateau. This plateau is called the Nasca Plains. On top of this plateau are geolyphs depicting insects, animals and humans, along with straight lines going north to south. This was discovered by Peruvian archaeologist Toribio Mejia Yesspe in 1927. The curious thing about these etchings in the ground is that the shapes of these insects, animals and humans can only be seen from an airplane...*high aloft!* The largest of these etchings is about 660 feet across. Carbon dating puts these geolyphs around 400 to 600 AD. Why they were made, no one seems to know. Anthropologist's speculate that it was for some religious reason to honor the gods.

**Speculation:** Although this area would make an excellent runway for any type of air vehicle, it more than likely served in giving direction to an airborne craft headed to another location such as Lake Titicaca. The carving of these geolyphs had to be directed from a great height because of their size otherwise they would not have made so precise.

## *Fact: Lake Titicaca*

Lake Titicaca sits on the border of Southern Peru and Northern Bolivia. It is about 110 miles long and 35 miles wide. Its depth starts at 100 feet and goes to 900 feet. It sits 12,500 feet above sea level making it the highest navigable lake in the world. The name of Titicaca was given to the lake by the ancient Indian culture that populated that area and roughly means panther leaping on rabbit. No one knew the reason why they gave that name to this lake until we entered into the space age. When space capsules with astronauts started circling our planet they discover that the lake indeed appears to take the shape of a panther leaping onto a rabbit.

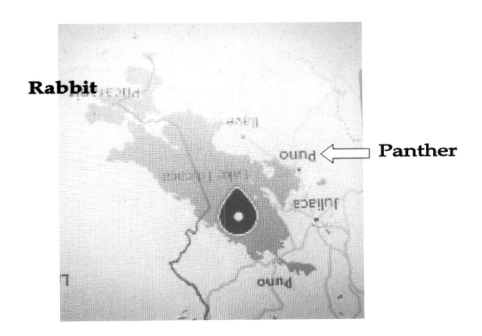

**Question:** *How would the ancient Indians have known this if the shape of this lake is only visible from the stratosphere? (It is interesting that some skeptics have laughably suggested that the inhabitants used primitive balloons to view this. My question is how did they dream-up and make the spacesuits that were necessary for lack of oxygen and extreme cold at such higher altitudes to view it?)*

**Answer:** It is quite evident that someone who saw the lake from a high altitude relayed that information, or had taken a prominent member of the tribe, perhaps the chief or head priest/shaman into space to show them the lake…*from a space vehicle!*

## *Rumor: The Sphinx*

There is a rumor that the written scrolls, which contain the knowledge of Atlantis is buried in a chamber beneath the Sphinx. Doctor Mark Lehner of Yale University's Department of Egyptology has done 37 years of field research on the Giza Plateau and the Sphinx. He has discovered unique information on this monolithic structure.

The Sphinx stands 66 feet in height and is approximately 240 feet in length. It was carved out of one mass of limestone and is one of the oldest if not the oldest of the monolithic statues on the face of this planet. The original name of the Sphinx is unknown as there is no mention of this monument on any of the tombs on the Giza

Plateau. It is believed that the Pharaoh Khafre ordered its construction as a guardian of the gateway to the underworld because limestone was collected from around the Sphinx to build his temple. However this presents some special problems as there is erosion at the base from rain water and drying periods indicating climatic change.

Approximately 9,000 BC, this section of land was a fertile plain at that time which was producing great quantities of rainfall. Approximately 7,000 BC, settlements appeared, which thrived until a climatic change occurred turning this land into a desert.

The problem that arises from this is that the Sphinx is estimated to be built well before 4,500 BC in a tropical environment because it is constructed from weathered limestone. Thus, this would mean that the construction of this monolithic structure before the reign of Pharaoh Khafre. It is also rumored that in the original construction a lion's head was a top this monolith and not that of a human head.

The Sphinx was carved out of limestone and eroded by water, not by sand or wind as the other structures in that area were found to be.

**Example:** Rain erosion causes surfaces to become round where as wind and sand erosion cause the surface of a structure to become square as seen in picture below.

Dr. Robert Schock, an expert on rock erosion, states that wind and sand erosion wears down limestone giving it square edges not round edges. The Sphinx is the only structure in all of the Sahara with round edges; all other structures have square edges.

Dr, John Kutzbich, a Paleoclimatologist (*individual that researches weather from previous eras*) from the University of Wisconsin, states that the Sahara was a fertile plain 9,000 years ago and had frequent rainfall and lakes. Approximately 8,000 years ago there was a climatic change in the Sahara which became its present state.

**NOTE:** There are some Egyptian Egyptologist that will deny these facts presented above in an effort to keep Egyptian discoveries credited only to Egyptian Egyptologists.

**Question:** *Did Pharaoh Khafre discover the Sphinx and utilize it for his own personal agenda? Did you know that there is another Sphinx on the planet Mars found by Nasa's Pathfinder Mission?* **(Photograph comparison following)**

**Question:** *Who designed the Sphinx? Was it the Atlantians, since its age is estimated to be older than 8,500 BC? It was previous assumed that mankind was still living in caves or possible only capable of building thatched huts. How then did these ancient humans get the technology to build the Sphinx?*

The final evidence is that the Sphinx's head is disproportional smaller than its body. The above information was found in a TV Special called "The Mystery of the Sphinx". **(See Illustration Below.)**

## *Speculation: About the Sphinx*

The Sphinx was carved from the natural limestone on the Giza Plateau known as the Mokkatam Formation, which was once part on an ancient inland sea. Being a much older structure, the Sphinx was converted into *Pharaoh Khafre's* temple as his engineers labored and changed the features on the head to resemble *Pharaoh Khafre*. This was indicated by the discovery of ancient tools around the base of the Sphinx.

## *Fact: The Pyramids*

Another interesting facet for discussion is the ancient pyramids. The Egyptians began building pyramids around 2700 BC. Curiously, in the great pyramid, four shafts were discovered that point to four galaxies in the night sky: Orion, Cereus, and the two Polar Stars. The last pyramid was built in Egypt about 1,000 BC.

In the America's the Incas constructed their pyramids around 1200 BC and ending their pyramid building during the Spanish conquest in 1519 AD. Interestingly enough the first Egyptian pyramid was a step pyramid built by King Josef, which is similar to that constructed by the Incas years later. However; there was no contact between these two societies. Also all of these pyramids appear to be astrologically *aliened to the same points.*

**Question:** *Why would their construction be so similar and what explanation can be given as to why were they astrologically aliened?*

## *Fact: The Inca Engraved Stones*

The engraved stones from the village of Ica in Peru, is a case of controversy in that if they are authentic, they may prove that an advanced civilization existed on earth hundreds of thousands of years ago. Some of these stones were found near the Nasca Plains. Ica is situated 150-200 kilometers from the village of Nazca.

It was here that Doctor Cabrera Darquea was first given one of the stones. He was intrigue by the carving on the stone and began to purchase more of them.

It was revealed to him that a man named Don Mendoza first discovered these stones in the village of Ocucaje. By the oxidation in the engravings of these stones it indicated that mankind had walked this planet longer than originally believed.

The stones were originally discovered embedded in saltpeter and ranged from 6 inches to a length of 2 yards. All stones appear to be highly polished and coated with some sort of wax to preserve the engravings. It is reported that these engravings show mankind co-existing with dinosaurs. Some of the stones are said to depict complex medical procedures as well. It has been estimated that there has been 50,000 of these stones found at that location. These stones show accurate depictions of Tyrannosaurs Rex, Triceratops, Stegosaurus and other dinosaurs along with medical transplants, blood transfusions, and other advanced technology etched on them. Organized by subject matter, the stone library also includes the races of man, ancient animals, lost continents, and the knowledge of a global catastrophe.

The stones were analyzed in the late 1960's and early 1970's *by a mining engineer* from the University of Bonn in Germany. Tests were also conducted at the Universidad Nacional De Ingeneira mining facility and analysis showed that the stones were from the Mesozoic era about 230 million years ago. A book titled "Forbidden Archeology" by Michael Cremo and Richard Thompson suggest that dinosaurs and man were around at the same time.

## *Fact: The Rock*

There is a rock that was discovered near a river in Glenrose, Texas by Roland T. Byrd, which is part of what is now known as Dinosaur Park. This rock was formed by cooling lava and has a dinosaur track embedded in it. There appears nothing unusual about a dinosaur track in a rock, however alongside the dinosaur track is clearly a humanoid footprint. Experts say that this is not a hoax because of carbon dating which showed both tracks to be made during the same time period.

Carbon dating is the rate of decay measuring the amount of radioactive Carbon-14 that's contained within that deposit. Although rocks cannot be carbon dated the deposit on which they exist can be carbon dated. Therefore, both tracks had to be made by stepping on this rock during the cooling process of the lava. Scientist cannot explain this.

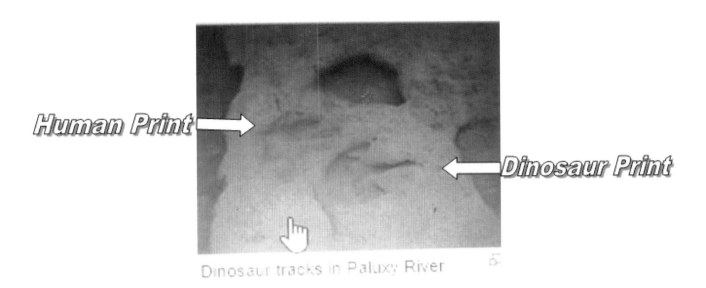

**(A Photograph with a wider view taken below of each set of tracks.)**

**Question:** *If this is true, how is it possible that two life forms separated by 65 million years can leave a foot print in the same rock at almost the same time?*

**NOTE:** In Chichanicia, Peru there are Mayan Paintings depicting dinosaurs and humans together, there is also found at a temple in northern Cambodia carvings of Stegosaurus.

## *Fact: The Star Child*

In the Mayan Region of South America a small misshapen child's skull of approximate 11 years of age was discovered over ten years ago. It was deemed the star child because of its unique features. The ancient Indians of that region would sometimes tie a newborn child skull in order to distort the child skull to honor their gods. Therefore, at first this discovery was not deemed to be of major significance. However, anthropologist decided to perform some test on the skull and the results proved quite interesting.

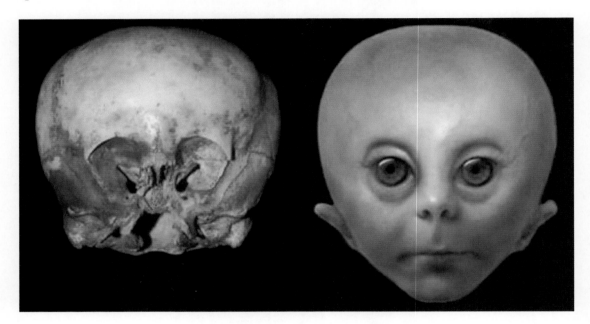

They first poured sand into a comparable skull of another child about the same age. Then they pour the contents of that skull into that of the star child's skull. The star child's skull held one third more sand than the normal skull, which meant that the brain capacity may have been larger than a normal human brain.

Next, they made an exact copy of the skull out of resin and sent it off to Hollywood special effects lab for re-construction of the child's features as is sometimes done for a crime scene. Incredible as it may seem, when the re-engineering of the face was completed it bore a striking resemblance to the creatures that were described in the Roswell Crash and in the abductions of Travis Walton and Betty and Barney Hill which was revealed during their hypnosis session and will be discussed later in this book.

It was both human and alien in appearance with large slanted eyes. Finally, because of the reconstructed results, they submitted the skull through DNA testing. Once more the results were surprising. The DNA showed that the mother of the child was definitely human; however; *the father was not human and of unknown origin.*

**NOTE:** *Pharaoh Akhenaten and his wife Queen Nefertiti also had similar elongated heads, but were tall and lanky in stature. And they too were said to be the union of "Gods" and human mortals.*

**Speculation:** It has been reported in some abduction cases that the aliens were more concern with reproduction of our species than with our intelligence.

**Question:** *Is it possible that humans are being used for reproductive experiments to aid an alien world in their reproduction in order to repair the damage to their population which may have been caused by some catastrophe?*

## *Fact: The Maps*

Unlike the arctic circle, which is comprised only of ice, Antarctica has land mass under its frozen surface. During the age of the dinosaurs this land was situated in a tropical region. It was a central part of a super continent known as Gondwanaland in the Mesozoic era. Then due to the moving of the under seas plates the land broke off and drifted to its present location thus becoming a frozen wasteland. Antarctica was not discovered until the early 1800's by a Russian naval officer and explorer Fabian von Bellingshausen. The exact land mass was eventually plotted by our satellite imaging system since it was covered by ice and could not be seen by the human eye. However, it is rumored that recently ancient maps were discovered showing the exact shape of the land mass without the ice that surrounds it. Two of which are named the Piri Reis map (1513) and the Oronteus Fineus map.

**NOTE:** The Oronteus Finaeus map, which is believed published in 1531, shows Antarctica as an "ice-free" continent with rivers, valleys, and coastlines, as well as the approximate location of the *physical* south pole, itself, by providing the correct longitudinal coordinates. It was drawn by the French cartographer Oronce Fine.

**The Piri Reis map (1513) shows Antarctica free of ice.**

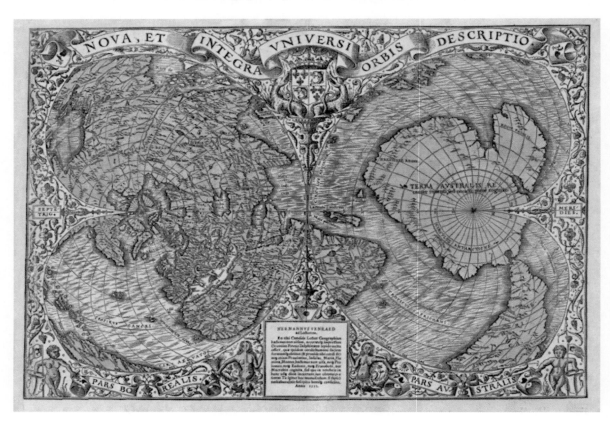

**The Oronteus Finaeus map (Oronteus Fineus map) shows Antarctica free of ice.**

**NOTE:** The cooling of the sun (*reduction of sunspot activity*) is known as a "*minimum*." The Maunder minimum coincided with the coldest part of the "Little Ice Age" (1500–1850). In England, the Thames river froze over. The Baltic Sea was covered in ice. We have just started another "*minimum*" right now, which means we're heading into at least another "Little Ice Age." Tell me, what does that do for anyone using "*Climate Change*" for an excuse to alter our way of life or to change human behavior? ***The sun can't be controlled by us, so I would think they had some other agenda.***

The Piri Reis map was found in Topkapi Palace in Istanbul in 1929 while excavating to create a museum: it is dated to have been created in 1513. This map shows the accurate and continuous coastline of South America with the exact contours of Antarctica. Professor Charles H. Hapgood and Mathematician Richard W. Starchen made comparisons with modern day photographs of the area taken from satellites and found that the map matches up precisely with these photographs. Thus the outline of Antarctica had to be made from an extremely high altitude.

It is further said that this map was copied *from Greek originals*, which could have been seen in the Library of Alexandria. It shows features of Antarctica without ice covering it drawn several thousand years before its discovery. However, critics claim that the reason why Antarctica appears on the map is that it is really showing part of South America and was curved because the individual that created the map could not get it on the paper properly. This map is covered in two earlier books: Chariots of the Gods by Erich von Daniken in 1969 and Fingerprint of the Gods by Graham Hancock in 1995.

**Question:** *Since the land has been covered with ice from at least the Cenozoic Era (100 million years ago) how, who and when was this map compiled?*

# Chapter III: The Modern Era of UFOs Sightings

In this segment we will endeavor to chronologically list major sightings of UFOs to show how they have been visiting our planet in our modern era. Through this we hope to thoroughly demonstrate that these visitations are not just rare occurrences, but appear on a regular basis. What makes these incidents creditable is the fact that they are transcribed in legal correspondence as newspapers, journals, documentaries and television news or editorials...*as well as releases by the military.*

## *The First Recorded Accounts*

**In 1884, the first account** that was noted or written down was by E W Reilley, a government inspector for agriculture and livestock. It was reported to him by Rancher John Ellis along with three other ranch hands that witnessed a fiery craft streak overhead to finally strike the earth. E W Reilley later went to the crash site and later reported it to the Nebraska Journal in 1887.

**1897, in Aurora Texas** an egg shaped UFO streaked across the sky witnessed by a great deal of the towns people before it hit a wind mill and crashed to earth. Many of the local inhabitants rushed to the crash site only to find a small dead alien being with a large head. They had a minister say last rites over the body before burying him in the local cemetery. Parts of the craft was thrown down a well. The owner of the property later was stricken by a severe case of arthritis from water from the property's well where some of the wreckage had been dumped and later removed, to be sealed up with a concrete slab. The alien body was later moved to an unmarked grave due to the bad publicity and out of town visitors that plagued the town.

**During 1941, in Cape Gerard, Missouri,** a Baptist Minister was said to perform burial rites for three aliens who died when their Disc shaped craft crashed. It was said that the beings were small with large heads and eyes. The minister was called from a nearby town at that time by police to give the three alien visitors "*Last Rites*." This story became a part of his own *death-bed testimony, which he swore was the absolute truth.*

## *Fact: The Battle for Los Angeles*

**In 1942,** World War II was in its early stages and on the West Coast of the United States rumors spread abundantly about the possibility of a Japanese invasion showing up one day in the aftermath of the bombing of Pearl Harbor on December 7, 1941. Homeland defense was on full alert and every evening searchlights relentlessly roamed the night skies. Then on a Tuesday night, February 24th, 1942, an oil refinery was fired upon by a surfaced Japanese submarine 90 miles from Los Angeles, which put the homeland military on alert. Tuesday night through Wednesday morning, February 25th, 1942 (the estimate time of 19:18 hours to 07:21 hours), outside of Los Angeles, California, a group of unidentified flying objects was observed in the night sky.

A total of twenty-five (25) Unidentified Flying Objects were either sighted visually or tracked by radar for approximately 12 hours. Thinking that the Japanese were invading, the entire coastline went on alert and the army reserves were immediately sent out to defend our West Coast. At least a half a dozen searchlights were focused on one disc in the sky, which was illuminated by these lights. Radar detection ranged these "aircrafts" at altitudes from a few thousand feet to as high as 20,000 feet moving at speeds varying from "very slow" to over 200 mph. Artillery and machineguns were set up to defend against the supposed invaders. **(A copy of the Actual Photograph is shown below.)**

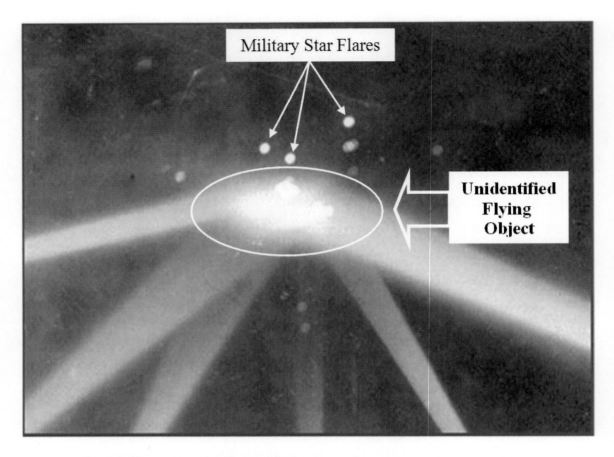

**This famous photograph above dubbed the "Battle of L. A." shown here, _is the original unaltered photograph taken that very night_.**

At 03:16 hours, elements of the 37th Coastal Artillery Brigade as well as other army units commenced firing upon one of these unknown aircrafts and a barrage of ammunition poured at the unidentified object in the sky for almost 15 minutes. Over 1,440 rounds of 12.8 pound anti-aircraft shells were expended as well as an undetermined amounts of other ordnance (such as 20 mm cannon, 0.50 and 0.30 caliber, machinegun, and rifle ammunition), but not one of these unknown "aircrafts" were brought down.

Only one of these *saucer-shaped* objects was photographed in the beams of eight searchlights. Then the objects flew up and out of sight. The only damage that appeared to be done was from the shrapnel that spewed the area and destroyed some of the local structures also causing the wounding of civilians plus killing 6 civilians as the metal pieces of shrapnel fell across the countryside raining for over an hour. This event had over a million witnesses.

After WWII had ended, the United States Government examined Japanese records and found there was no indication that these UFOs were ever from Japan. The United States Government stated in later years that it was a weather balloon and the military unit that launched it was afraid to admit to it because of the disturbance it caused. However, General George Marshall in a memo to President Roosevelt stated although the UFOs might have been Japanese commercial aircraft, it did imply *they resisted every attempt to shoot them down and were completely unknown.*

**Question:** *If the accurate artillery and machinegun fire could not bring this object down then what was it?*

**Speculation:** A television show called "Fact or Faked", whose premise was to investigate unknown phenomenon, tried to recreate this occurrence with the aid of the army reserves. First they thought that the spotlights crossing each other would make it appear that a solid object was in the sky. Thus using the same type of spotlights of that time period and in the same configuration, they pointed them at the same spot in the night sky. However, the lights went through each other and gave no indication of a solid mass that was seen in a photograph, which was taken that night in 1942 during the incident. (See Previous Page) Next they thought that it might have been a blimp (although there was no record of any blimps being launched that day). Thus, they sent up a high altitude balloon and had the army reserves fire at it. The balloon blew apart in seconds. Therefore, it could not have been a balloon. Again, no resolution was made as to what the object was, other than that it was *definitely* a solid mass that was being observed...*and resisted every effort to bring it down.*

**Question:** *What did cause the pandemonium that night? Could it possibly have been a real UFO?*

Following is the actually General George Marshall memo to President Roosevelt about the "Battle of L.A.)

WAR DEPARTMENT
OFFICE OF THE CHIEF OF STAFF
WASHINGTON

February 26, 1942.

MEMORANDUM FOR THE PRESIDENT:

The following is the information we have from GHQ at this moment regarding the air alarm over Los Angeles of yesterday morning:

"From details available at this hour:

"1. Unidentified airplanes, other than American Army or Navy planes, were probably over Los Angeles, and were fired on by elements of the 37th CA Brigade (AA) between 3:12 and 4:15 AM. These units expended 1430 rounds of ammunition.

"2. As many as fifteen airplanes may have been involved, flying at various speeds from what is officially reported as being 'very slow' to as much as 200 MPH and at elevations from 9000 to 18000 feet.

"3. No bombs were dropped.

"4. No casualties among our troops.

"5. No planes were shot down.

"6. No American Army or Navy planes were in action.

"Investigation continuing. It seems reasonable to conclude that if unidentified airplanes were involved they may have been from commercial sources, operated by enemy agents for purposes of spreading alarm, disclosing location of antiaircraft positions, and slowing production through blackout. Such conclusion is supported by varying speed of operation and the fact that no bombs were dropped."

Chief of Staff.

Franklin D. Roosevelt Library
DECLASSIFIED
OD DIR. 5200.9 (9/27/58)
Date- 3-10-59
Signature- Carl L. Spicer

## *1943 - 1944; During World War II*

Balls of light were observed by American, British, Japanese, and German pilots following and or pacing their airplanes. These balls of light became known as *Foo Fighters*. They appeared in several colors as follows: Red, Orange, White, Yellow and Green. At first they were thought to be *new* German enemy craft, however; later after the war was over documents were found that the German pilots thought they were American surveillance aircraft.

When these balls of light came close to any aircraft they would cause the instruments of the aircraft to fail or malfunction. These orbs could not be detected on radar and appeared to be under some sort of intelligent control. They could out perform any aircraft at that time with almost impossible maneuvers. They were often seen while dogfights were occurring as if they were taking accounts of our aircraft performance. Later when jet aircraft came into use, these orbs seemed to have no affect on our jet engines turbulence.

**A World War II photograph of "*Foo Fighters*" flying alongside military aircraft. The "arrows" you see in the photo had been placed in the photograph on previous page by military intelligence people of that time period. The aircraft appears to be Japanese (Ki-48) bombers.**

**A Kawasaki Ki-48 in profile.**

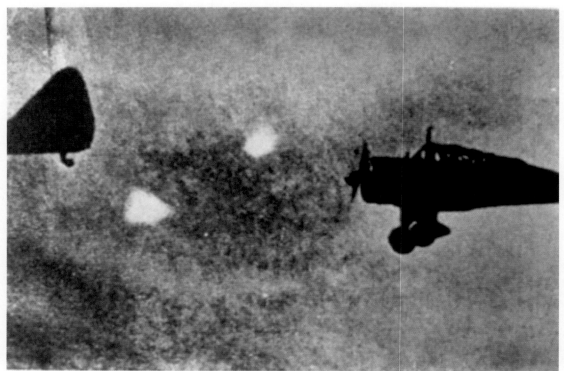

**More "Foo Fighters" in a formation with British Aircraft.**

**British Aircraft in WW II photo believed to be Westland Lysander variants.**

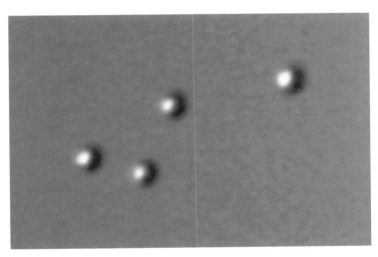

**Photograph taken from an aircraft of "*Flying Orbs*"**

**NOTE:** Orbs have been seen and video recorded with a ***film camera*** making crop circles in an English farmer's field. This film was thoroughly scrutinized, and begrudgingly yielded the conclusion: <u>*there was no computer generation performed on the film whatsoever*</u> and that <u>*the film was completely authentic*</u>. *Below are six frames taken from that very film.*

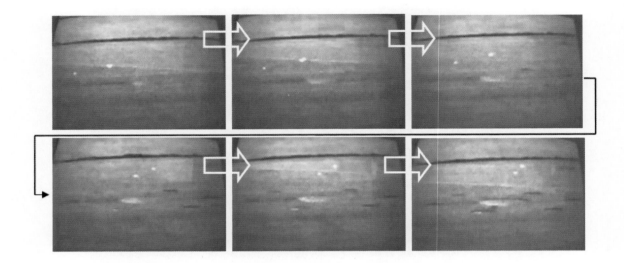

**NOTE:** Looking for their 15 minutes of fame and glory, a few unstable people have come forward claiming that they had made these crop circles. However, there is a little problem with their confessions, crop circles have been made for hundreds of years...on different continents and in different countries,..*even before the advent of balloons, airships and airplanes*. These confessors would then have to have a life span of over 500 years to do so. *Do the skeptics and experts find that the confessors' explanation really plausible?*

**ADVICE:** *To the debunkers, skeptics and other such experts*: "It's better to keep your mouth shut and appear stupid than open it and remove all doubt" - **Mark Twain**. In other words, keep an open mind *and do the research before condemning what other individuals are stating that they personally observed.*

### NASA Mission STS 51A

It should also be noted that Astronaut James McDivit stated that on his Gemini-4 mission he had photograph an orb. Orbs were also seen on Apollo-12 and 17 missions. They were also seen on Discovery shuttle STS-51A in 1984 and shuttle mission STS-37 in 1991. It was ascertain that they were not ice fragments or space debris and they appeared under intelligent movement.

## *June, 1947; Captain Smith*

Captain Smith while piloting a DC-3 along with his copilot and stewardess did observe a UFO near Mt. Adams and Mt. Rainier in Washington State. It is possible that this was one of the UFO's that Ken Arnold reported which would also give creditability to his observation.

## *June 21, 1947; Maury Island, Puget Sound, Washington State*

Harold Dahl, Fred Crisman and one other crewman were on a fishing boat salvaging logs with Dahl's young son when they said they saw six flying donut-shaped discs (each about 100 feet in diameter) appear above them over the water, off of the south shore of Maury Island over Puget Sound in Washington State. One of the discs appeared to be in trouble as it shuttered and tipped dangerously close to the water. The presumed malfunctioning craft then dropped what appeared to be tons of a steaming hot molten substance (slag) in the water, some onto his boat, and some onto the nearby beach. As the story goes, the heat and debris killed Dahl's dog, and burned and broke his son's arm. Some objects dropped onto their boat resembled a substance that was like lava or "white metal."

**NOTE:** It is Standard Operating Procedure (SOP) for any aircraft having a serious malfunction to dump fuel if a crash seems imminent. With less fuel aboard, the aircraft becomes lighter and more controllable, and the risk of fire is reduced along with possibility of explosion. If the aircraft is forced down, it raises the probability of surviving and escaping in a controlled crash.

The ejection of the metal seemingly fixed the problem with the malfunctioning craft, which was joined by the other five vehicles and sped away. Some of the material was then gathered up from the boat, water, and the beach. The three men and teenage boy then made a report to the local police.

Now here is where it gets strange. About a week or two later, Harold Dahl met with Kenneth Arnold (at the request of the United States Army Air Force) and was interviewed about the UFO encounter. Arnold (now considered the premier UFO investigator) thought the incident to be of high interest to national security, and recommended the slag material should be analyzed by the United States Army Air Force.

Two Army Air Force officers (both pilots) visited Harold Dahl several weeks afterwards, on August 1st. One of these men was Frank Brown, who not only served as an intelligence officer, but also worked in the field of counter-intelligence. The two officers listened to the tale and took a statement from Harold Dahl along with the box filled with the slag pieces of various sizes gather for evidence.

The two officers brought the evidence directly to their aircraft, a B-25 bomber at McCord Air Field. The medium bomber had been completely overhauled two weeks prior, so there should have been no problems with the aircraft whatsoever.

The investigators took off to return to Hamilton Army Air Force Base in California. However within the first twenty minutes of the flight, the B-25 suddenly suffered from unknown difficulties and crashed, killing both officers and destroying whatever evidence they were carrying. The FBI closed the case without any resolution. Although the debris covered a large track of forest in Washington, material recovered at the crash site (although inconclusive) was deemed to be "highly anomalous," as so stated in a report.

**Newspapers from August 1, 1947 describe the crash of the Air Force B-25 bomber.**

**Fact & Speculation:** This story has been *meticulously* retold by the "UFO Hunters." And as they demonstrated, there might have been a possibility that the slag may have consisted of ferrous material (such as magnetized iron…that was perhaps…***combustible to electrical spark***). But we'll never know for sure since all of the players in this drama are now all deceased, and the slag…lost in time. However, we still can speculate.

This material may have disrupted the electrical system onboard the aircraft, which may have accidentally shorted the bomber's internal 125 Amp electrical system by the material somehow, which then may have ignited and caused a horrific onboard fire. What leads to this presumption is the fire itself. It burned far hotter than 1220°F (the melting point of aluminum).

**NOTE:** ***Thermite is a pyrotechnic composition of metal powder fuel and metal oxide.*** When ignited by heat, spark, or flame, thermite undergoes an exothermic oxidation-reduction reaction. Most varieties are not explosive, but can create brief bursts of high temperature in a small area.

Thermites have diverse compositions. **Fuels include aluminum (aluminum is common because of its high boiling point)**, magnesium, titanium, zinc, silicon, and boron. Oxidizers include boron oxide, silicon oxide, **iron oxide**, copper oxide, chromium oxide, manganese oxide, and lead oxide.

**NOTE:** Magnetism was first seen in a mineral form called Lodestone. Lodestone consists of an **iron oxide compound,** a combination of both iron and oxygen molecules. Therefore, with all the components needed for such a fire in very close proximity of each other,...**and all materials out-gassing their individual molecules** in a minute vapor (which everything does do) within the closed compartment of an airframe, all it took was a single spark (probably from a spark-gap capacitor, which would have been typical in the electronic equipment of that era) to cause ignition...or detonation.

**NOTE:** How a spark-gap capacitor works is that the capacitor has two exposed electrodes of opposite polarities separated by a small premeasured gap of air to act as both an insulator and trigger for the capacitor. The gap of air is decided by the needed breakdown voltage between the two opposing electrodes. When the voltage between the electrodes reaches the critical breakdown of the air's insulating properties, a spark jumps from one electrode to the other to temporarily bring them both to the same electrical potential, thus neutralizing an overload condition. Usually, this is to ground level.

**NOTE:** I have personally worked on electronic equipment of either from that era or a more modernized version of it, which still could have been seen as late as the mid 1980s. For example, the old Cathode Ray Tube type television sets (until plasma and LED types replaced them) retained spark-gap capacitors for the 20,000 volt power supply need to create the picture on the screen of the Cathode Ray Tube TV set. Such capacitors were to ensure the voltage did not surge to the point of exploding the tube and killing its viewers. However, if there were flammable vapors or fumes within the household or dwelling where the TV was situated, it could ignite those fumes very easily, causing either a fire or an explosion by its discharge. And this would have been how the aircraft caught fire and exploded from the intermingling from the very out-gassing of the incompatible materials it carried.

Although this fire should not have been encountered in flight under normal...or even most abnormal circumstances, however; there was never a guarantee of that.

This fire was verified by soil samples at the crash site along with **the presence of aluminum oxide on aircraft parts**, showing that the aluminum was subjected to very hot, very intense heat. And with the deaths of these two army officers began the long line of tragic loss of pilots and aircrews attempting to solve the riddle of UFOs.

**Speculation:** Now, with the facts of this story totally revealed, let's do a little postulating. Let's say everything reported is true. We have a "donut-shaped" circular craft dumping its fuel of molten slag (fuel because they dumped a lot of it) to stay aloft. Slag has been used to term things like molten metal...and lava...like in the Earth. So now we have several possible clues as to the principles used by this (for better purposes) *spaceship*.

Think of it in terms of the Earth. Lava is molten liquid made of the elements (in various amounts) silicon, oxygen, aluminum, iron, magnesium, calcium, sodium, potassium, phosphorus, and titanium (plus other elements in very small concentrations). It moves beneath our feet creating a magnet field that protects us from the solar winds and provides us with direction finding (using a compass of course). The Earth's interior of lava also creates our world's gravity (although this is assumed based mostly on mass, but there are other factors involved here as well).

The inner core of the Earth is made of superhot iron (this is also assumed), but to our knowledge, molten iron is not magnetic in itself. However, the pressure is so great that this superhot iron crystallizes into a solid (which makes it magnetic). Convection caused by heat radiating from the core, along with the rotation of the Earth, forces the liquid iron in the outer core and mantle to move in a rotational pattern. Therefore, it is believed, that these rotational forces in these layers lead to magnetic forces around the axis of spin.

So that means there is another process involved in creating the Earth's magnetic fields. And the processes (at least for the Earth) are a pressurized inner core supplemented by rotation of the outer core and mantle. So in order to create a magnetic field using molten slag, we would have to rotate it as it flowed around a magnetic center.

Now, let's say the fuel slag is heated to a molten state and is made to spin while it rotates in a circular flow around the outline of the donut-shaped spaceship, using the elements of a hydraulic system. What types of fields might be created then by this slag movement? Gravitational and magnetic fields might be created of course in varying degrees of intensity, either one or the other, or a combination of both.

Now we can make a somewhat accurate assertion that the craft's propulsion is by gravitational and magnetic fields developed by moving certain molten materials in a rotating circular pattern surrounding a magnetic center. And with the right materials and designed system (using the Earth as a model), we can attempt to construct a prototype by a little reverse-engineering and find out how exactly the donut saucer works.

What comes to mind is to pressurized molten material and flow it through a *rifled* hydraulic line (capable of withstanding 2800°F) to create the proper flow pattern over and around a centered tungsten cable (tungsten has a melting point of 6192°F while iron has a melting point of 2800°F) passing electric current to generate the magnetic field. And to get the different maneuvers being performed by the craft (to make it go up, down, right and left), we could have three such systems in our craft: one at the top, one at the bottom, and one in the middle towards the skin of the craft. The response time could be varied in microseconds by changing the amount of current in each individual or all of the three systems. This just might give us the propulsion effects we are looking for (like turning at right angles at high speeds). Not too hard, was that? At least it gives us a starting point to develop a prototype of a vehicle that could possibly carry us to the stars.

And we, ourselves, have created similar devices. Have you ever hear of an electromagnetic propelled submarine? A 10 to15 foot model was successful tested in 1968, I believed, by Cambridge University students as a science project and has been experimented on by different navies ever since. They even made a movie about one, *The Hunt For Red October*. And today, many others (like the University of Washington) are continuing development of such a system.

There is also Professor Subrata Roy of the University of Florida, who believes he can build (what he dubbed) the saucer WEAV, or wingless electromagnetic air vehicle. Using his plasma research experience, Roy developed a craft with no moving parts such as engines or propellers. Instead the surface of the craft is studded with electrodes to ionize the air, creating plasma. Electric currents are sent through the plasma to give it lift. Actually, it just might work!

## *On June 24, 1947; Kenneth Arnold*

Kenneth Arnold while on a search and rescue mission near Mount Rainier, Washington (State) for a downed aircraft described a series of crescent-shaped objects (classified as Delta Shaped now) flying north of Mt. Rainier, at estimate speed he calculated to be from 12,000 to 18,000 miles per hour. He reported it immediately to the FAA (Federal Aviation Administration) saying that they made no noise. The news media picked up this report and from Arnold's description of this account likening their movement to saucers skipping on water, the term "Flying Saucers" was coined, and thus UFOs captured the interest and imagination of the entire world.

**Kenneth Arnold displaying an art rendering of the UFOs he saw.**

Arnold's sighting was actually corroborated by a prospector by the name of Fred Johnson on Mt. Adams. He wrote AAF intelligence (Army Air Force – before its re-inception as the United States Air Force) that he saw six of the objects on June 24, which he viewed through a small telescope a little bit *after* the time that Arnold did, providing an almost identical description of these craft. He said they were "*round and tapered sharply to a point in the head and in an oval shape.*" He further noted that the objects *created an electromagnetic disturbance causing the rotation of the needle on his compass to oscillate.*

An evaluation of the witness by AAF intelligence at the time found him to be quite credible. Ironically, Johnson's account was listed as the first unexplained UFO report in Air Force files, while Arnold's was *dismissed* as a mirage and later as a flight of Pelicans despite the fact that Johnson had described a continuation of the same event as viewed by Arnold. (Arnold was regarded as a highly skilled and experienced pilot, with over 9,000 hours total flying time, half of which were devoted to Search and Rescue, and as a *medical transport* Mercy Flyer.

## *July 3 – 4, 1947; an object crashed in White Sands, New Mexico known as the Roswell Crash*

**The location of the crash was the Foster Ranch outside of Roswell, New Mexico. We have dedicated an entire chapter to this historic incident following this section because of the amount of material contained within our research.**

### Fact: January 7, 1948; The First Pilot Lost in Pursuit of a UFO

**Photograph of a P-51D, Type of Fighter Aircraft flown by Capt. Mantell**

Captain Thomas F. Mantell of the Kentucky Air National Guard had logged a total of 2,167 flight hours during his military career. In World War II he had been awarded the Distinguished Flying Cross and the Airman's Medal for Heroism with 3 Oak Leaf Clusters (*which means that he was awarded this medal 4 times*). He flew in the invasion of Normandy to support the landings of allied troops. (The majority of his flight time, however, was spent in transporting paratroopers, special operations, and cargo.) On January 7, 1948 this 25-year old pilot would display his courage again, but this time his name would go down in history as being synonymous with the chasing of Unidentified Flying Objects.

**NOTE:** Thomas Mantell Jr. was born in Franklin, Kentucky, June 30, 1922. He was a graduate of Male High School, in Louisville. On June 16, 1942, he joined the Army Air Corps, graduating Flight School on June 30, 1943. During World War II, Mantell was assigned to the 440th Troop Carrier Group, 96th Troop Carrier Squadron, 9th Air Force. He was awarded Distinguished Flying Cross, and Air Medal w/3OLCs for heroism. Following the war he returned to Louisville, joining the newly organized Kentucky Air National Guard, as Flight Leader, "C" Flight, 165th Fighter Squadron on February 16, 1947.

It began at 13:20 hours (1:20 p.m.) when the Kentucky State Police called Godman Army Airfield at Fort Knox with reports of a unusual flying object *about 300 feet in diameter* over Maysville, Kentucky. The military police then contacted the control tower where it was relayed to Technical Sergeant Blackwell. After numerous calls from civilians to the airfield, the sighting was reported to the Base Commander, Colonel Hix.

Colonel Hix then went to the control tower to observe the situation with binoculars. Hix along with other military personnel watched the unknown flying object for over an hour in the southwest sky. Other military bases were also observing this same aerial phenomenon. Clinton County Army Airfield described the object as very white and one quarter the size of the moon. Some of the ground personnel stated that it appeared to be stationary at times or slowly drifting. This latter sighting by the ground personnel may have been the planet Venus.

At 14:50 hours (2:50 p.m.), Captain Thomas Mantell and his wingmen Lieutenant Albert Clemmons and Lieutenant B.A. Hammond along with a fourth pilot were ferrying their P-51 Mustangs from Marietta Army Airfield in Georgia to Stanford Army Airfield in Louisville, Kentucky when they were contacted and asked to investigate the situation. It was at this time that the fourth pilot stated that he was low on fuel and continued on to the original destination.

Flying to the designated area, they intercepted the unknown at 15,000 feet approximately 15:10 hours (3:10 p.m.). It was then that Captain Mantell called in, stating that they had a visual on the bogey (unknown), which was above them and that it was *a gigantic metallic object.* Captain Mantell said that it was moving at half their speed and that he was going up to get a better look at it.

**NOTE:** When Mantell led the way in the climb to 15,000 feet, upon reaching that altitude, he radioed the following statement back to the control tower:

"The object is directly ahead of and above me now, moving at about half my speed…. It appears to be a metallic object or possibly reflection of Sun from a metallic object, **and it is of tremendous size**,...I'm still climbing,...I'm trying to close in for a better look."

**NOTE:** Army regulations state that oxygen masks must be used at altitudes above 10,000 feet otherwise those pilots would begin to suffer from Hypoxia (altitude sickness) from loss of Oxygen, which can cause unconsciousness and death.

Only Lieutenant Clemmons had an oxygen mask on his aircraft. The three pilots climbed to 22,500 feet when Lieutenant Clemmons and Lieutenant Hammond began feeling the effects of dizziness and advised Captain Mantell that they were breaking off pursuit. Captain Mantell said that he was going to continue upwards to 25,000 feet to get a better view of the object. Whether he made it to that altitude is debatable, but we assume he did. And Mantell's plane was observed by an eyewitness, Glen Mays, who lived near Franklin. **Mays stated categorically that Mantell's plane exploded in midair.**

*"The plane circled three times, like the pilot didn't know where he was going," reported Mays, and then started down into a dive from about 20,000 feet. About halfway down, there was a terrific explosion."*

Captain Mantell was discovered dead in his P-51, *the cockpit still sealed*, by the Franklin, Kentucky fire department. The fighter had crashed on a farm just south of the town. His watch had stopped at 15:18 hours. The Unidentified Flying Object had vanished from sight around 15:50 hours.

Adding to all this, was the testimony of Richard T. Miller, who was in the Operations Room of Scott Air Force Base in Belleville, Illinois. He made several profound statements regarding the crash. While monitoring the radio communications between Mantell and Godman tower, he swore he heard this statement very clearly: *"My God, I see people in this thing!"*

At a briefing on the morning after the crash, Miller said the investigators had stated that Mantell died *"pursuing an intelligently controlled unidentified flying object."* What further complicated the matter was the sighting at Lockbourne Army Air Field in Ohio. At 19:15 to 19:30 hours, personnel stated that a huge object came very close to the ground, and then after hovering for a few moments, climbed at a speed in excess of 500 MPH before leveling off at around 10,000 feet. It was further stated that this object made three complete turns circling the tower (*including against the wind*) and appeared larger than a C-47 aircraft. It then moved off in a south-southwest direction and could not be a balloon as the wind was blowing from east to west that day. The military personnel that viewed this object were made to sign a report and told that this incident was classified and that they were not to discuss it.

**NOTE:** One of the soldiers at Lockbourne Army Air Field in Ohio made this observation, *"Just before leaving it came to very near the ground, staying down for about ten seconds, then climbed at a very fast rate back to its original altitude, 10,000 feet, leveling off and disappearing into the overcast heading 120 degrees. Its speed was greater than 500 mph in level flight."*

Captain Mantell's crash was investigated by the Army Air Force's research group for UFO paranormal activities called Project Sign (which later became known as Project Grudge). Their first determination stated that Captain Mantell had mistaken Venus for a UFO and that he had passed out somewhere between 25,000 and 30,000 feet due to lack of oxygen. They later changed their conclusions to **a Project Skyhook balloon** because Venus would have been almost invisible due to the brightness of daylight during that time period, which would only have allowed Venus to be viewed a half hour at any one location. They also stated that if there was any haze, Venus would not have been seen at all...*unless one was searching for it*. In those conditions, Venus would have been hard to distinguish from the rest of the sky, *which made casual detection highly unlikely*. However, Venus was visually easier to see from the air than rather from the ground.

Also, there is the testimony of Godman Base Commander Guy F. Hix, who stated to reporters that he observed the craft for almost an hour through binoculars. He would not have confused what he saw with the planet Venus.

For their next dismissive theory to try and give plausibility to the sighting, the Army Air Force jumped to the Project Skyhook balloon. As discussed in Chapter IV, this balloon was supposed to gather radiation data and was made of reflective aluminum and polyethylene. The balloon approximate size was 100 feet and was capable of reaching an altitude from 60,000 to 75,000 feet. Several of these balloons were said to have been launched that day; **but there is absolutely no proof of that.**

The UFO was also said to have traveled from Madisonville, Kentucky to Fort Knox, Kentucky a distance of 90 miles in 30 minutes *at altitudes considerably lower than the jet stream*. This would mean that the wind speed over Kentucky had to be 180 miles per hour or *at hurricane force*. However, *since there was no hurricane that day*, the Air Force researchers then reverted to a combination of both theories stating that the balloon and Venus combined together to cause the UFO reports.

**Speculation:** Although there is some question as to exactly how Captain Mantell died, however, the more important question as to what he was chasing still remains the biggest mystery. Some researchers said that Captain Mantell's body was not in his plane when it was found and that the government is covering up this information. They said that he was abducted by the aliens, however; this theory is discounted due to fact that the cockpit was sealed. *Suspiciously, his remains were buried within 24 hours of the crash.* There were numerous statements from the ground personnel that witnessed the event, *but supposedly reported statements* from both wingmen Lieutenant Clemmons or Lieutenant Hammond with regards to what happened seem more than questionable and do not correspond with what Captain Mantell reported and what they said they saw.

**Question:** *There appears to be no documented direct interview with either Lieutenants Hammond or Clemmons to be found anywhere. The question is...why?*

**Question:** *Why did the government alter their reports? The existence of Project Skyhook was well known because they were the ones who promoted the project in the media. Or perhaps was there something more highly classified than a simply weather balloon?*

**Question:** *Does the Air Force still consider Project Skyhook classified after 70 years and after much media publicity starting in 1947?*

**Question:** *Has the Air Force also classified Venus as Above Top Secret too? If so, what about the Earth and Mars? (Yes, we are being just as ridiculous as the United States Air Force.)*

It was said that it was a Skyhook Balloon that may have been what Captain Mantell chased and that several were launched that same day. However, as stated previously, *there were absolutely no records* of any Skyhook balloons or for that matter…any balloons *at all* being launched that day. The only balloon that was launched was on the day before, January 6, 1948 at 0800 hours from Camp Ripley in Minnesota. Captain Mantell last report at 15,000 feet was that the unknown was a gigantic metallic object. What did the other pilots see then if it wasn't a balloon?

If Captain Mantell was close enough to claim that it was *"of tremendous size"* and metallic then he should have been close enough to recognize that it was a silver shaped balloon. Even if the color of the balloon was not the normal color used, Captain Mantell should have still identified the object as a balloon being a pilot with 2,167 flight hours. A balloon is still a balloon. On that day, the planet Venus was barely visible due to the brightness of the sun and you had to be looking in the right location to view Venus. This would hardly qualify as being either *"of tremendous size"* or metallic.

**NOTE:** The P-51 Mustang has a top speed of 437 MPH, a service ceiling of 41,900 Feet, a range of 1,000 Miles, and could reach its service ceiling in approximately 11 minutes. This fighter aircraft, the P-51, was considered the premier aircraft of World War II, it also flew combat missions in the Korean War, and made a brief appearance in the Vietnam War as an attack aircraft, the Piper PA-48 Enforcer, due to its ruggedness and high reliability.

**The Piper PA-48 Enforcer of the Vietnam War**

How could a Skyhook balloon with a wind speed of 15 MPH (the highest speed recorded that day) outrun an aircraft that was specifically designed for pursuit and interception? Witnesses at Lockbourne Airfield 4 hours after Captain Mantell's crash stated that an unknown flying object came near to the ground and then took off at a speed in excess of 500 miles per hour (which could out-run a P-51). This would discount both Venus and the Skyhook balloon for neither qualifies with these reported multiple accounts. Calculating the size of the Skyhook balloon in ratio against the distance of Captain Mantell's P-51 it would have been almost impossible for him to have seen a Skyhook balloon at 15,000 feet if the balloon was at its minimum altitude of 60,000 feet. Refer to the Calculations below:

**NOTE:** Linear perspective is a perfectly inversely proportional relationship. The formula to find the height of an object is:

$$h = a / d$$

*or to find how big it should look to you from your position, then use this inverse formula:*

$$a = h \times d$$

whereas "h" is the apparent height, "d" is the distance of the object, and "a" is the actual size of the object. So if you want to find the true height of an object in the distance, multiply the apparent height with the distance the object is from you. And since we already know the diameter of a weather balloon is 100 feet, we can use that to solve for the apparent height that Captain Mantell saw in his P51...*if it was a weather balloon.*

1 Mile = 5,280 feet

So if it was a weather balloon Captain Mantell saw, at one mile, the balloon would appear as 100/5280 feet in size or 0.189 foot = 2.29 inches in diameter to Mantell.

At 45,000 feet away from the balloon (8.5 miles), the balloon would appear as 0.0022 foot or 0.026 inch in diameter (1/40th of an inch).

At 60,000 feet away from the balloon (11.36 miles), the balloon would appear as 0.00167 foot or 0.020 inch in diameter (1/50th of an inch).

This would hardly be considered as *"of tremendous size."* Unless Captain Mantell had the eyesight of an eagle, he would only have seen a mere speck of reflective light from the balloon or may not have been able to see the balloon at all. Since the sun's bright light almost obscured the planet Venus, which was supposedly too appeared to be ¼ the size of the moon and was at a 6 degree arch in the sky, it too may not have been visible to Captain Mantell according to astronomers that reviewed the case. Then neither of the fore mentioned objects would have appeared gigantic or metallic at 15,000 feet.

**Question:** *What of the observation at Lockbourne Air field after Captain Mantell's crash?*

Since when has the planet Venus ever landed on an airstrip then take off again at 500 miles per hour? The object on the airstrip was said to be 1,200 feet in diameter, which clearly eliminated the planet Venus or a Skyhook balloon.

**Question:** *If Captain Mantell was not in pursuit of Venus or the Skyhook balloon, then what were he and the other pilots in pursuit of?*

**NOTE:** Although America had operational jet fighters at that time, the P-59 and the P-80, only the P-80 had a maximum speed (599 MPH) a little over 100 miles faster than the Mustang, but it was restricted to approximately the same service ceiling. **(Chart following)**

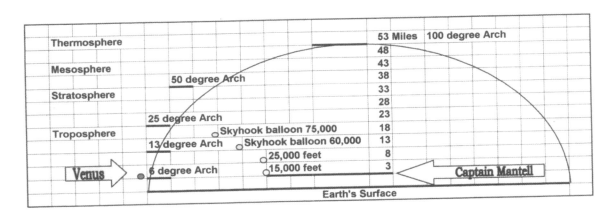

**Inches to Miles to Degrees of Arch**

3 ½ = 53 miles = 100 degree arch divided by 2

1 3/4 = 26.5 miles = 50 degree arch divided by 2

7/8 = 13 miles = 25 degree arch divided by 2

7/16 = 6.5 miles = 12 degree arch divided by 2

7/32 = 3 miles = 6 degree arch

By looking at the graph at a 6 degree Arch aligns the Planet Venus almost exactly ahead of or just below Captain Mantell's altitude, yet he stated he was going up to get a closer look at the UFO. If he were in pursuit of the Planet Venus (which was at a 6 degree arch or 12,500 feet to 15,000 feet), he would have stayed at the same altitude and not climbed to 25,000 feet. By the calculations on the previous page the Skyhook balloon would be too small to be considered gigantic and metallic as he reported to the tower. Therefore, the only plausible deduction would be that Captain Mantell was in pursuit of the same UFO that was seen at Lockbourne Army Airfield.

*But there is more to this story: the crash debris from Mantell's aircraft!* Additional information has come to light by Captain James F. Duesler, who was one of several military officers at Godman, who came forward in 1997. At that time, he was retired and living in England. Duesler stated that he and several other officers **actually saw the gigantic UFO hovering over Godman Airfield that day!**

Duesler, who was a pilot and crash investigator, stated, "The UFO was a strange, gray-looking object, which looked like a rotating inverted ice cream cone."

Shortly after the crash, Duesler visited the site of the downed P51, and made these observations, "The wings and tail section had broken off on impact with the ground, and were a short distance from the plane," he recalled. "There was no damage to the surrounding trees and *it was obvious that there had been no forward or sideways motion when the plane had come down.*

"There was very little damaged to the fuselage, which was in one piece, and no signs of blood whatsoever in the cockpit. There was no scratching on the body of the fuselage to indicate any forward movement and the propeller blade bore no telltale scratch marks to show it had been rotating at the time of impact, and one blade had been embedded into the ground.

"The damage pattern was not consistent with an aircraft of this type crashing at high speed into the ground. (*It was like the aircraft had been lifted up by a crane and then suddenly dropped to the ground.*) Because of the large engine in the nose of the plane, it would come down nose first and hit the ground at an angle. Even if it had managed to glide in, it would have cut a swath through the trees and a channel into the ground. None of these signs were present."All indications were that it had just belly-flopped into the clearing. I must admit, I found this very strange." **(A single photograph followed by a photo montage of Captain Mantell's crashed aircraft are shown below.)**

**Photograph 1 of Wreckage of Captain Mantell's P-51**

**Photograph Montage of Wreckage of Captain Mantell's P-51**

**Question:** *If Captain Mantell's airplane was seen to explode around 10,000 feet by numerous witnesses, why wasn't there any sign of fire and explosion in the inspected wreckage? And why wasn't the debris field scattered over a much wider area due to an explosion?*

This means that Captain Mantell's P-51 didn't actually explode...*but only looked like it did!* How was that possible? Could it possibly been some type of advanced weapon or device employed by the UFO that caused that phenomenon? If so, were the aliens really seeking to kill Captain Mantell...or perhaps...save him? We'll never know. However, what we do know is that *every piece of debris from Mantell's P-51 was removed (even the smallest pieces vacuumed up) and classified by the Army Air Force as Top Secret,...**to this day,**...but why?*

**NOTE:** The UFO theory had received even more credence from reports filed the very next day after Mantell's death. On January 8th, residents of Clinton, North Carolina, reported a cone-shaped object moving through the skies at incredible speeds. I would hardly call this a coincidence.

## *July 24, 1948; Eastern Airlines Flight 576*

Eastern Airlines Flight 576 observed an elongated craft that whizzed by their plane at approximately 700 miles per hour. The craft was shaped like a cigar and approximately 100 feet in diameter and had two rows of windows. Captain Clarence Chiles and 1st Officer John Whitard stated they were 20 miles south of Montgomery, Alaska when this sighting occurred.

## *SWORN TESTIMONY - KOREAN WAR*

**(Over 30 instances of pilots, and far more numerous ground and sea observations were made by airmen, soldiers and sailors, respectively, during the Korean War. If all the listing of all these experiences were compiled, it would generate a hefty book in itself. Only two have been detailed here to give the reader an idea of how vast the number of sightings were reported...*or eventually reported.*)**

**Air Incident: September, 1950; -Source: NICAP/Blue Book - Pilot: Unnamed.**

A U.S. Navy plane on a combat mission was approached by two large discs. The aircraft's radar was jammed, and its radio transmitter was blocked by a buzzing noise each time a new frequency was tried.

**Testimony:** "Very early in the morning, three fighter-bombers took off from the flight deck of a U.S. aircraft carrier. The sun hadn't risen and there was bite in the air. Two men -- pilot and a radar gunner -- occupied each of three planes assigned to a routine mission, bombing and strafing a truck convoy that combat intelligence expected to be winding along the floor of a valley about a hundred miles from the Yalu river. The takeoff was routine, as was the flight to the target area. At about 7 A.M., just as the sun was breaking above the mountains in the East, the aircraft were proceeding North, ten thousand feet above the valley floor. Radar observers had their eyes peeled for the target.

"I was watching the ground below for the convoy, reported...and was startled to see two large circular shadows coming along the ground from the Northwest at a high rate of speed. We were flying North above a valley which was surrounded on the East and West by mountains, with a pass directly ahead of us to the North. When I saw the shadows I looked up and saw the objects which were causing them. They were huge. I knew that as soon as I looked at my radar screen. They were also going at a good clip—about 1000 or 1200 miles per hour. My radar display indicated one and a half miles between the objects and our planes when the objects suddenly seemed to halt, back up and begin a jittering, or fibrillating motion. My first reaction, of course, was to shoot. I readied my guns, which automatically readied the gun cameras. When I readied the guns, however, the radar went haywire. The screen bloomed and became very bright. I tried to reduce the brightness by turning down the sensitivity, but this had no effect. I realized my radar had been jammed and was useless. I then called the carrier, using the code name. I said the code name twice, and my receiver was out—blocked by a strange buzzing noise. I tried two other frequencies, but couldn't get through. Each time I switched frequencies the band was clear for a moment, then the buzzing began.

"While this was going on the objects were still jittering out there ahead of us, maintaining our speed. About the time I gave up trying to radio the carrier, the things began maneuvering around our planes, circling above and below. I got a good look at them. I had never seen anything like them before, and I learned after we reached our carrier that the other men in that flight had the same opinion. They were huge. Before my radar set was put out of commission, I used the indicated range plus points of reference on the canopy to determine their size. They were at least 600 or possibly 700 feet in diameter.

"The objects had a silvered mirror appearance, with a reddish glow surrounding them. They were shaped somewhat like a coolie's hat, with oblong ports from which emanated a copper-green colored light, which gradually shifted to pale pastel-colored lights and back to the copper-green again. Above the ports was a shimmering red ring, which encircled the top portion.

"When the things maneuvered above us, we saw the bottoms of them. In the middle of the underside was a circular area, coal black and non-reflective. It was simply inky black, and it is important to note that although the whole object jittered while maneuvering, the black circular portion on the bottom was steady and showed no indication of movement.

"When the objects seemingly finished their inspection of the Navy planes, they took off in the same direction from which they had come, and disappeared at a high rate of speed."

**Ground Incident: May, 1951, near Chorwon, North Korea (60 miles north of Seoul)**

Interview given January, 1987 by former PFC Francis P. Wall to John Timmerman, an associate of the J. Allen Hynek Center for UFO Studies (CUFOS) in Chicago, Illinois.

Preparing to bombard a nearby village with artillery, soldiers of the 25th Division, 27th Regiment, 2nd Battalion, 'Easy' Company observed a strange sight up in the hills—like "a jack-o-lantern come wafting down across the mountain." It is night, and they were located on the slopes of a mountain above an abandoned Korean village they were bombarding.

The light continued on down to the village to where the artillery air bursts were exploding. It had an orange glow in the beginning as it stood among the artillery airburst for forty-five minutes to an hour, yet remaining unharmed. But then this object approached the soldiers, while turning to a blue-green brilliant pulsating light.

Lt. Evans, company commander, then ordered all the men under his command to fire upon this object with their M-1 rifles using armor-piercing bullets. The bullets did hit their mark, but the soldiers heard their rounds bounce "in pings" off the object's metal skin. However, the object went momentarily wild, and its light was going on and off. The light then went fully off completely, but briefly. The object began moving erratically from side to side as though it might crash to the ground. (Up until this point, the object made no sound whatsoever.)

Suddenly the soldiers heard a sound like that of a diesel locomotives revving up. The soldiers were then attacked by some form of a ray that emitted in pulses, which swept in waves they could visually see only when it was aiming directly at them. It was compared to how a searchlight acts when it sweeps around, The soldiers then felt a burning, tingling sensation all over their bodies, as though something were penetrating them.

Lt. Evans ordered his entire company into bunkers, and there they waited unknowing what would happen next. The object hovered for a while, lit up the whole area with its light, and then shot off at a 45 degree angle and was gone.

Three days later the entire company of men had to be evacuated by ambulance. Roads had to be cut in there to haul them out. Virtually the entire company was too weak to walk. It was found they had dysentery and subsequently, an extremely high white blood cell count, which the doctors could not account for. No report was ever made of the incident by any of these soldiers out of fear reprisal by the United States Army.

### *July 9, 1952; Two Pan American Pilots*

Two Pan American Pilots saw a UFO up close as it flew by their airplane. It was being chased by a F-94 Interceptor near Washington D.C. This was the start of the Washington D.C. flap.

### *July 19, 1952; Washington D.C. (July 19, to July 20, 1952):*

**NOTE:** Actually, there were a series of unidentified flying object reports from July 12 to July 29, 1952, over Washington, D.C. July 19th and 20th were the most publicized incidents, however.

Edward Nugent, an air traffic controller at Washington National Airport (now renamed Ronald Reagan Washington National Airport), was the first one to spot seven unknown objects on his radar,15 miles south-southwest of the city at 11:40 pm. These objects were not following any established flight paths, nor were any of these aircraft suppose to be in that area. Nugent immediately brought these aircraft to the attention of his supervisor, Harry Barnes, a senior air-traffic controller at the airport. Both men then watched the unknowns on Nugent's radarscope.

Harry Barnes later wrote: "We knew immediately that a very strange situation existed…their movements were completely radical compared to those of ordinary aircraft."

Barnes then had two other controllers check Nugent's radar, which they confirmed was working properly. Barnes then immediately called National Airport's other radar center and the controller stationed there, Howard Cocklin. Cocklin told Barnes that he also had the objects on his radarscope as well. Howard Cocklin, however, at times was looking out of the tower window and could see one bright orange object at about 3,000 feet.

At this point, more objects appeared in all sectors of the radarscope. Finally, they moved over the White House and the United States Capitol. Barnes then frantically called Andrews Air Force Base, located 10 miles from National Airport. A flight of fighter jets were then immediately scrambled to bring down these intruders. However, these Unidentified Flying Objects out-maneuvered and out-distanced the U.S. jets in mere seconds with an acceleration of speeds *in access of 7,000 MPH,* which were picked up by radar, promptly vanishing into the sky. With nothing now found on radar, the fighters left the area and headed back to base. But eerily a few minutes after the fighters had landed,…*the UFOs returned.* At this time a passenger plane was taking off and saw 9 white lights flashing past them and then disappeared.

Finally at 05:30 a.m. in the morning, the drama ended. Around this time, E.W. Chambers, a civilian radio engineer living in the suburbs of Washington D.C., observed "five huge disks circling in a loose formation. They tilted upward and left on a steep ascent." Near the same time, the *unknowns* vanished from radar.

The following Saturday at 11:30 p.m., eleven to blips appeared on radar again at Andrews Air Force Base. One jet fighter was dispatched to investigate. Upon reaching the target area, he was surrounded by four UFOs. The pilot became nervous and radioed the tower asking what he should do. No one had a answer for him and shortly later the UFOs sped away. Later the Air Force attributed the sightings as a freak weather disturbance.

**(Below are copies of two photographs taken of that night.)**

(Below a copy of a front page of the newspaper "Washington Post" marking the event.)

## May 18, 1953; A UFO Crashed

A UFO crashed and was found two days on May 20th near Kingman, Nevada nuclear test site at least 20 miles from route 66. In Arizona, just prior to the crash, several of the local inhabitants stated that they witnessed 8 disc flying in a erratic dog fight type maneuver and 3 were seen to be crashing. The military sent a team out from Creech Air Force Base. One of the team members was Arthur Stazsel from the Air Material Office of Special Studies. He stated that they found one downed craft with 3 alien bodies dressed in silver suits. These aliens were about 4 feet in height with tan skin and large eyes (*remember the star child from a previous chapter*). The other two craft were destroyed beyond recognition.

## Rumor: Presidential Visitation

On February 20, 1954; according to President Eisenhower's great granddaughter, Laura; President Eisenhower went missing for 12 hours at Edwards Air Force Base for emergency dental work, but instead was told later he was engaged in a meeting with blue-eyed Scandinavian in appearance aliens. It was said that a treaty was made, which stated that they would give us technology in exchange for humans needed in genetic research.

## Fact: 1955; Landing at Edwards A.F.B. – Gordon Cooper

In 1955, Astronaut Gordon Cooper (before joining the space program) was a pilot stationed at Edwards Air Force Base. While on temporary assignment in which he supervised the filming of F-86 fighters landing, Gordon Cooper observed a saucer-like craft hovering over a nearby dry lake bed. This metallic object then extended three pods and landed in the lake bed. The film crew immediately turned the camera on the craft and started recording this event. The craft only remained at this location for a minute or two before taking off at an incredible speed and disappearing into the sky. This footage was developed the next day, but no one was permitted to view it. A courier then picked up the film that same day and took it to Washington D.C. It was never seen or heard about again.

### 1955; Captain William Coleman

Captain William Coleman was in the process of delivering a bomber from an air base in Florida to an air base in Mississippi when he spotted a disc shaped object on their flight path. He reported it to his superiors and it was sent to Project Blue Book were it was dismissed and not investigated.

### On November 3- 4, 1957; in Levelland, Texas

In Levelland , Texas an egg shaped UFO was observed two days in a row over White Sands testing grounds.

### May 20, 1957; in Monstown, England

In Monstown, England, U.S. Pilot Milton Torres was ordered to shoot down a giant UFO while flying from an English air base at the height of the Cold War. Torres was ordered to open fire on a massive UFO that lit up his radar, according to an account published by Britain's National Archives. The fighter pilot said he was ordered to fire a full salvo of rockets (*a full salvo of rockets which consisted of 24 air to air missiles*) at the UFO moving erratically over the North Sea. However, at the last minute the object picked up enormous speed and disappeared.

### On December 6, 1958; Over the Gulf of Mexico

Over the Gulf of Mexico a B-58 bomber was returning to its home base when it encounter a mother UFO craft with 3 small shuttle craft following. It was reported that these UFOs were traveling at 5000 miles per hour as they flew past the bomber.

A United States Air Force B-58 "Hustler" Bomber

**On September 19, 1961; in Portsmith, New Hampshire, Betty and Barney Hill were abducted. See Abductions Chapter** of this manuscript.

## On April 24, 1964; in Sornoro, New Mexico

In Sornoro, New Mexico; Patrolman Lonnie Zamora was on patrol at night when he spied an egg shaped object in the distance in the desert. Upon arriving at the area where the craft had landed, he found 3 small figures with large heads standing by the UFO. Upon seeing him they hurried into the craft and took off. The figures were dressed in white coveralls and left 4 inch footprints in the sand. He reported the incident and the United States Air Force sent investigators over to the landing site. They tried to match the craft's prints to that of known aircraft, but could find no match. The government investigation was logged in Blue book as unexplained.

## Rumor: Landing at Holloman A.F.B.

**Supposedly sometime in 1964- 1965:** Paul Shartle was security manager and chief of requirements for audio and visual programs at Norton Air Force Base. His job was to categorize and log all films made by and for the United States Air Force. One day he received a directive to create a film about Unidentified Flying Objects and Alien Life Forms. He was told to contact a movie producer by the name of Robert Emenegger, who was already working on this project for the government. They met and began to work on the script. In their research they viewed a film from Holloman Air Force Base in White

Sands, New Mexico that was made in 1964 or 1965 or (actually, there have been a couple of dates when this is claimed to have happened). It showed three disc shaped craft circling above the air base runway. Eventually, one of these discs came down to the runway while the other two continued circling the base.

**NOTE:** This is a procedure used by the US Military in combat situations when picking up or dropping military personnel in a hostile environment and is performed by helicopter pilots. One helicopter would land to pickup or drop off the troops while the other two would circle to give immediate air support against any ground enemy units should it be necessary. This operation is performed until all personnel from all three helicopters are retrieved or dropped off.

The disc that hovered above the runway began to land as three pods emerged from the bottom of the craft. Once on the ground, a seamless door opened on the craft and three humanoid type aliens emerged from the craft. These humanoids had an odd grey complexion with large noses. It was at this point that the Base Commander came out and greeted them.

**NOTE:** A handful of UFO researchers over the years have mentioned these 'Large-Nosed' Aliens in various books, seminars, etc. Some speculated they come from Betelgeuse in the constellation of Orion. Some speculate that a treaty was signed with these visitors and that three (or more) humans left the earth with them. However, the Betelgeuse Star is a M2 Spectral star and out of the normal star range to support life. This researcher believes that these aliens are from Zeta 2 Reticuli, which is suppose to contain a race identical to our own. Since Zeta 1 and Zeta 2 Reticuli are both G 2 star systems, it is more plausible that this is the star system where the tall aliens originated from.

*(See **Travis Walton** article in this manuscript under abductions)*

Paul Shartle and Robert Emenegger inquired about this film and were told that it was theatrical footage. Paul Shartle said that he knew nothing about this film and if it was theatrical footage, it would have fallen under his jurisdiction and have been catalog. They asked if they could use this footage for the ending to their production and they were told no. This film was sent off to Washington D.C. Later it was learned that when the filming of this landing occurred, one of the airman believed to have work in security became disturbed by what he had seen and upon his day off, went to town. There he began talking about the incident. After returning to the air base, he was never seen again. It was then that Paul Shartle and Robert Emenegger received a call from a government representative and was informed that the project had been cancelled.

**Speculation:** It appears that someone in the government wanted to release information about Unidentified Flying Objects and Alien Visitations, but at the last minute the higher authorities decided against it.

**NOTE:** For the longest time there were rumors about such a film, which would discuss alien life on another planet that was to be produced by the government entitled "Cosmic Journey" that never premiered. This particular information was revealed on the television special "UFO Cover Up Live", however, the story of the Holloman Air Force Base Landing started circulating throughout the security community the day after the incident occurred, which is when this author became aware of it.

## *On August 3, 1965; Outside of Santa Ana California*

Outside of Santa Ana California a man named Rex Heflin who was a Highway Maintenance Engineer for the Orange County road department was doing a survey at the crossroads of Myford Road and Walnut avenue when out of the corner of his eye he spotted a silver disc in the sky. Reaching into the back of his car he pulled out a 38-year old Polaroid camera which he used for his job. He then snapped four pictures of the strange object through the windshield of his vehicle. The craft was hovering at about 150 feet. It then tilted and flew off to the right across Myford Road. His radio was not functioning , thus upon his return to his office he notified his supervisor. **(Below is a copy of two of the photographs with a "Blowup.")**

These photos have been examined for any tampering, but were determined to be authentic. Computer enhancement and other analysis was later conducted by Robert Nathan at the Jet Propulsion Laboratory, working with first-generation prints and copy negatives made by the newspaper. Among other things, the analysis established photographic evidence to confirm the (reported) "light beam" on the underside of the object. However, the United States Air Force in a statement at that time, declared the photos to be a hoax. (*Please do remember, these are the same people who also **indirectly implied** that "the planet Venus" landed and took off again at Lockbourne Airfield in Kentucky, 4 hours **after Captain Thomas F. Mantell's P-51 fighter crashed.** That does wonders for their credibility, doesn't it?*)

Another little twist to this story was that Heflin had turned over three of the four originals to a man (or two men, the stories differ) who claimed that he represented the North American Air Defense Command (NORAD). However, NORAD denied that they had ever sent out an investigator or had the slightest interest in the photos. Yet, the original photos were mysteriously returned in 1997 (*after 32 years of being missing*).

One final note to this story was that the now defunct TV program "UFO Hunters" dedicated a complete show on this UFO case, using state-of-the-art computer programs along the updated survey techniques and mathematics Heflin himself had used to identify the distance of the object and its size. The UFO Hunters completely validated and vindicated that the object and the photographs were quite real and exactly what Heflin had claimed.

### March 5, 1966; Minot Air Force Base

In the evening hours of that night, a UFO appeared over Minot Air Force Base *and 10 Minuteman missiles went into countdown to launch!* The Capsule Officers responsible for launching the missiles had to go into abort shutting down all power to the launch system. The red glowing UFO then left the area heading due south. A similar situation had occurred over a Russian missile site...*around the same time and date.*

### March 5, 1966; Lieutenant Colonel Salas of SAC

Lieutenant Colonel Salas of SAC (Strategic Air Command) was a Launch Controller Officer for nuclear Minuteman missiles. At approximately 23:00 hours (on Malxstrom Air Force Base in Montana), an alert for their security personnel was initiated due to the observation of a huge reddish orange glowing disc-shaped object in the sky near their nuclear facilities. This report was mentioned in the movie "Out of the Blue."

Accordingly, this strange craft affected the launch ability of 20 nuclear missiles on the base's weapons systems in its southward path. (Each of these weapons has an independent launching system.) However, every individual system was affected at the same time and all power was lost. The craft moved slowly across the night sky taking 30 minutes to cross from horizon to horizon. This caused 30 minutes of launch down-time, in which the missiles sat completely inoperative...and could not have responded to any military launch emergency.

### March 6, 1967; Security at Francis E Warren Air Force Base in Wyoming

At Francis E Warren Air Force Base in Wyoming, security was put on alert as all nuclear missile systems started failing around 01:30 hours. There were 14 flight complexes at this location scattered over a hundred miles. On a southward path each complex lost launch capability for approximately 30 minutes as a huge red object passed overhead heading southwards.

**Speculation:** This object appeared to be the same UFO that hovered over Minot and Malxstrom Air Force Bases that also were affected, with Francis E Warren Air Force Base in its path along with their their strategic areas. In plotting its southward path, this may have been heading for Nellis Air Force Base (Area 51) in Nevada. Therefore, as many as four Air Force Bases may have been affected by it. This would have been a serious threat to national security, and if true, would have been covered up at the highest levels of government. The UFO may have been giving this country a demonstration of their capability and/or warning about the use of nuclear weapons.

### *September, 1967; Chase over S-2, Offiut Air Force Base*

Early September 1967 – At approximately 13:50 hours, a silver disc was observed moving slowly from east to west by government security and maintenance team members over a high security area in a mountainous region known as Sierra 2 or S-2. It was a clear day with no clouds in the sky. This craft was observed by members of the military Missile Maintenance Team (MMT), Electrical Maintenance Team (EMT) and Target Alignment Team (TAT), at least 22 individuals in all.

One of these individual went down into the concrete support shelter and used the direct phone line to notify the launch control officers. They must have notified higher authorities because two jet fighters appeared from the east apparently in pursuit of this silver disc. The fighters may have been dispatched from Offiut Air Force Base in Nebraska. The disc was clearly visible to the naked eye appearing ¾ of an inch in diameter and ½ inch in height about 500 feet away thus was estimated to be 12 feet in height and 35 feet in diameter.

The craft then began a series of maneuvers as if to test these fighters' abilities. First it shot straight up and hovered as if waiting for the two fighters to catch up. The fighters circled upward to get the disc in their sights as the disc was now over their heads. This UFO then shot across the sky and as the fighters leveled off the disc dropped straight down. This caused the fighters to circle downward. When the two fighters leveled off again, the UFO shot straight up once more causing the fighters to complete a circle in their flight path. Once more the two fighters leveled off to get it in their sights and at that point the UFO shot straight upward and out of sight.

**NOTE:** This sighting is a result of first-hand information. The important aspect of this sighting was that it occurred in daylight an in clear weather observed by government security personnel over a highly restricted area. *(A craft resembling the object in the photograph below is what was described.)*

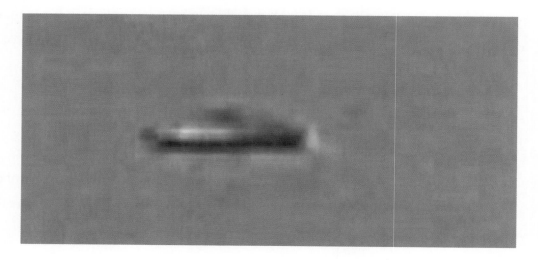

## *Fact: October, 1967; Shag Harbor, Nova Scotia*

On October 4, 1967, a ten year old boy named Christopher Styles saw a glowing strange orange light moving across Shag Harbor around 23:00 hours from his bedroom window. He darted from the house to get a better look at the light and found that it was clearly surrounding a solid object. At that same moment a young man named Laurie Wilkens and three of his friends were driving home and saw the same object moving across the sky. They decided to follow this unknown object, which appeared to be in route to Shag Harbor in Nova Scotia, Canada. Eighteen year old Norman Smith also at that time saw five lights in the night sky moving towards Shag Harbor making no sound. He notified his father about the sighting.

Eventually, the orange lit craft went into the waters of Shag Harbor. The people that had been following this light became alarmed thinking now that it had been an airplane and that it had crashed into the water. Laurie Wilkens along with eighteen other bystanders had watched the craft go under the water. Laurie Wilkens notified the Royal Canadian Mounted Police (RCMP). At first the police thought that it was a prank call. However, their switchboard became overly active with frantic phone calls about the craft hurling into the water. The Royal Canadian Mounted Police sent out two officers to investigate the reports. Norman Smith and his father followed the police cars to the harbor. Royal Canadian Mounted Police sent out an alert for anyone that had a large boat to join in the search for the downed aircraft. The officers assigned the investigation organized the volunteers and started a rescue operation at the site where the craft went down.

Two large fishing vessels were used along with several smaller boats to cruise the area of the sinking looking for survivors or debris. They found orange foam floating on the surface, which deteriorated when they tried to scoop it up. However, there were no signs of survivors or debris. In the meantime, the rescue center at Halifax, Nova Scotia checked for missing aircraft and for filed flight plans that would have taken an aircraft over or near the crash site. They found that no aircraft were schedule in this area that night. They then realized that they were searching for something other than a regular aircraft.

Coast Guard Cutter # 101 arrived at the scene of the crash on October 5 and patrolled the area looking for bodies and or debris until October 8, 1967. The Canadian military then declared the crash craft to be of a UFO. On October 6, they sent divers down to find the craft. Ray MacCloud, a newspaper reporter, began filming the search because the Canadian people wanted to know what crashed in the harbor and the Canadian government refused to release any information. An unnamed military diver did confide to Ray MacCloud that they did find something, but refused to elaborate on what it was.

In 1992, Christopher Styles launched his own investigation and began gathering documents released by the Freedom of Information Act. Being overwhelmed by the material, he then contacted Donald Ledger to help review the documents that he had received. These documents revealed that there had been numerous UFO reports on the night of October 4, 1967 all across Canada.

Two pilots had seen a UFO and an explosion that followed. The records of deceased Jesuit Priest (Father Burt Gatvey who was known outwardly to deny the existence of UFOs and Alien life) showed evidence that a UFO was seen in Shelbourn, Nova Scotia that night. Christopher Styles checked with divers that were at Shag Harbor. They told him that the UFO had moved from Shag Harbor to Shelbourn under the water. Once in Shelbourn, a second UFO was seen entering the water.

The United States and Canada both sent divers down and observed alien beings from the 2nd UFO assisting other beings from the 1st UFO making repairs. It was at about that time that a Russian submarine (which may have been monitoring the situation) was detected entering U.S. and Canadian waters. At that point the United States and Canadian vessels abandon monitoring the UFO to intercept the Russian submarine. It was after this that the two Unidentified Flying Objects emerged from the water and flew off.

**Speculation:** It appears from the above information that the United States government and the Canadian government validates defense minister Paul Hellyer statements and are working in conjunction with each other with regards to UFO incidents. The United States government appears to be the primary administrator in these operations and determines what information is to be released and what is to be suppressed from the general public. It also appears that these governments are using critics and clergy to down play, miss direct and or deny the existence of UFOs and extraterrestrial beings. It is apparent that the United States and the Canadian governments are aware of the existence of UFOs and are deliberately concealing this information from the public.

## On March 24, 1969; in Papua New Guinea

On March 24 1969, in Papua New Guinea, Australian Missionary Reverend William Gill along with three other individuals and 30 natives observed 4 small aliens on top of a saucer shaped craft. They waved to them and the aliens waved back. The craft hovered there for several hours before taking off.

## On July 20, 1969; Apollo 11 Astronaut Buzz Aldrin

Apollo 11 Astronaut Buzz Aldrin while on the Moon was reported saying to NASA "They're watching us," referring to two Disc overhead of the astronaut and landing site. Both Buzz Aldrin and Neil Armstrong also claimed to see structures on the far side of the moon. **(Below Possible image of what they may have seen?)**

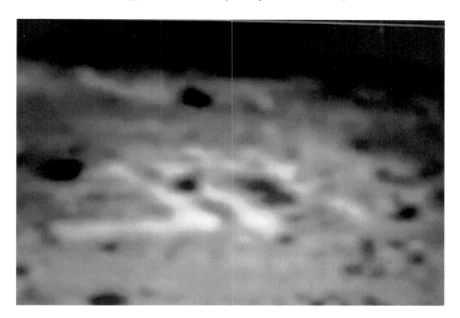

**(Alien installation on the moon?)**

## *February 5, 1971; Apollo 14*

Apollo 14 Astronaut Edgar Mitchell 6th Astronaut to walk on the moon reported to NASA that while on the moon, he did observe a UFO and 3 alien beings watching him.

## October 11, 1973; Pascagoula, Mississippi, Hiskson and Parker

See Chapter on Abductions in this manuscript.

## *Fact: Near Collision over Ohio*

October 18, 1973; 20:00 hours – Mansfield, Ohio – Lieutenant Colonel Coyne and his crew were returning to base when they became aware of a bright red object paralleling their UH1H helicopter. They were at an altitude of approximately 2,200 feet when they made the observation. The object then circled the helicopter and headed straight for them. Lieutenant Colonel Coyne and his crew thought at this point it was a missile. The object came so close that it almost hit their propeller blades. They put their helicopter into a power dive to avoid collision. Suddenly, it vanished. However, when they looked up, they found it was above them. Once more they attempted a steep dive. However, instead of going down they felt themselves being pulled upward. The needle on their altimeter went from 1,700 feet to 3,700 feet in almost a few minutes. Finally, the craft released the helicopter and flew off. Lieutenant Colonel Coyne in the documentary "UFOs Are Real" stated "that it had a higher degree of technology than we possess." A family of five that was driving home at the time, had pulled off the road to observe this encounter between the UFO and the helicopter. The UFO was estimated to have been 80 feet in length. Other members of the crew were Sergeant John Healey and Richard Jesse confirming the incident.

## *In May of 1974; Fort Hancock, New Jersey*

A Nike Hercules missile base was being shutdown and dismantled in agreement with the disarmament accords previously signed by the late President Nixon. During the middle of that month for three consecutive nights, the missile storage/launch area was visited by *unknown* aircraft of very unique design that hovered silently above the restricted area for as long as fifteen minutes at a time.

On the first two nights, the craft took on the appearance of a *spinning glowing egg* whose burning white radiance varied in luminosity. However, on the third night, a new craft appeared: a glowing triangle. The vehicle hung motionlessly in the air for a few minutes pointing in the direction of the base. Then after slowly turning clockwise around by about 120 degrees, it zipped away at an incredible rate of speed estimated to be *over several thousand miles an hour!*

No report was ever made by any of the one hundred plus soldiers of the United States Army who personally witnessed these events. Regardless that each man was a highly trained observer holding a high security clearance, no one dared to *officially* report the incident to the Army Air Defense Command. And for good reason…they feared what their own government would do to them if they ever did. Loss of pay, loss of security clearance, a wrecked military career, being held under guard in isolation for months at a time were just a few of the punishments associated with acknowledging the witnessed truth. So a vow of silence was begrudgingly taken by all, never to reveal what they saw over those three nights—until now. For you see...*I was one of those soldiers.*

*R. G. Risch: Author of Beyond Mars Crimson Fleet*

**Seen above: Nike Hercules Air Defense System, which had Nuclear Capabilities.**

**November 1975, Silgreaves National Forest, Travis Walton;** see Chapter on Abductions in this manuscript.

## Fact: February 5, 1977; Broad Haven Village

At a school in Broad Haven Village located in Wales, the children on that Friday afternoon reported seeing a landed silver cigar shaped UFO near their school faculty. They also reported seeing a being dressed in a silver suit and helmet near the craft. The principal decided to disregard the children's insistence as it was after hours. On the following Monday, the students still emphatically repeated the story that they saw a UFO behind the school. Finally the principal sat the children down and had them draw what they saw. *Stunningly*, all the drawings matched...*exactly!* Shortly afterwards, reports of UFO sightings started filtering in from witnesses in and around that area. There were so many reports that the Ministry of Defense sent their people to check it out since it was also near several major air bases.

## Fact: January 8, 1978; Alien Shot and Killed at McGuire Air Force Base

According to Air Force Maj. George Filer, retired, who was an intelligence officer at McGuire Air Force Base in the 1970s and Security Police Officer Jeffrey Morse, an alien 3 or 4 feet tall was seen near the fence line separating Fort Dix and McGuire AFB. A Fort Dix United States Army Military Police Officer spotted and confronted the being with a verbal challenged. However, the alien took off running. The MP then panicking, drew his weapon and hastily fired five rounds from his .45 into the being. However, the wounded alien managed to climb over the fence. The creature was found dead near a hangar by a runway apron on the McGuire side of the fence.

Two men found it: an Air Force security guard and a New Jersey state police officer, whom the guard had admitted onto the base after the shots were fired and a search ensued. It was said to have a large head and long arms. The OSI (Office of Special Investigation), the air force's investigation unit, took immediate charge of the investigation upon their arrival. It was said that the alien body gave off such a foul stench that it made anyone coming near it ill.

A C-141 flew in from Wright-Patterson Air Force Base in Dayton, Ohio, and took the alien body away. Accordingly, the MP who shot the alien was transferred to Guam, lost his rank, and was interned in a detention facility, where he underwent constant interrogation and brutal treatment by his guards. After over 4 months of his mistreatment and false imprisonment, the MP was released and immediately discharged from the United States Army. There was no court-martial or any other legal proceeding involved in the matter whatsoever. A call to the Air Force press office received this response: "We have no records that such an incident described took place on that date at Joint Base McGuire-Dix-Lakehurst."

**ADVICE**: Unfortunately for them, we do have the information from multiple sources that corroborate the incident did indeed "*took place on that date at the Joint Base*

*McGuire-Dix-Lakehurst*." Maybe you Air Force guys should update your Guard Report for this incident.

## *Fact: December 24 to 26, 1980; Rendlesham Forest / Bentwaters Sighting*

There have been a variety of theories on the Bentwaters / Rendlesham Forest sightings. Claims of the actual encounters with an unknown craft by the airmen have been muddied by repudiations made by self-serving critics armed with mistaken perception of circumstances, lack of reason, misinterpretation of known occurrences, and their theatrical hunches of hoaxes all in attempt to discredit the highly trained military witnesses of this event.

Therefore in reviewing all material available, we, the researchers have tried to piece together an accurate account of events that occurred during that 3-day period. How accurate our explanation is up to the reader for interpretation. It should be noted that two years prior to this incident, two triangle shaped UFOs were tracked on radar flying over the British side of the air base *at 4,000 miles per hour*. At that period in time our fastest aircraft was only capable of speed no faster than 1,000 miles per hour.

Rendlesham Forest lies 8 miles east of Ipswich village in Suffolk, England and on the perimeter of Bentwaters Air Force Base which is shared by the Woodbridge RAF (Royal Air Force of England) and the United States Air Force. This incident first came to light around 0300 hours on December 24, 1980 when strange lights were sighted by members of the United States Air Force Security Alert Team (SAT), which occurred near the east gate. The lights were said to have moved over to the weapons storage facility and beamed a blue light onto the area before moving towards the forest where it vanished. The SAT unit went to investigate and saw this light moving through the forest.

**Mockup of object seen by Staff Sergeant Penniston and Airman 1st Class Burroughs**

Staff Sergeant James Penniston and Airman 1st Class John Burroughs of the 81st Security Police Squadron along with Airman Edward Cabansao responded to the area of the incident. They estimated that they first saw the object about 160 feet away from them. It was egg or acorn shaped glowing yellow with red blinking lights surrounding it and a blue light at the bottom. The craft was hovering off the ground, however; there were three indentations in the dirt and broken branches around where it came down. Upon approaching the craft Staff Sergeant Penniston claims to have touched the craft as he circled, it noting that it was warm to the touch.

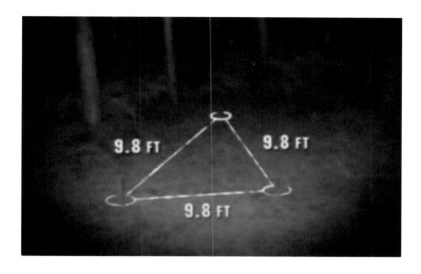

**Unidentified craft's landing imprints**

He pulled out his notebook and began to make a sketch of the object that was before him. He stated that the object remained there for 45 minutes as he made entries into his note book. Afterwards the craft ascended into the trees causing the three men to dive to the ground. Sergeant Penniston later stated that upon arriving home ,he took put his note book and began scribbling down the *1*s and *0*s in the book. He filled up 12 pages of these letters. It was later reviewed and turn out to be a Binary Code, which is the second basic code used in computers, of which Sergeant Penniston had no knowledge of. Computer programmers analyzed the code and it discovered it reference several geo-locations on this planet.

Penniston also sketched symbols located on the side of the craft in his note book.

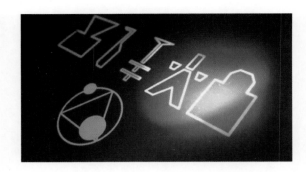

The following night during a family Christmas party, Deputy Base Commander Lieutenant-Colonel Charles Halt was informed that the UFO had returned. Wanting to put this incident to rest and thinking that this was most likely a misidentification of a known object, Lieutenant-Colonel Halt responded to the scene of the incident with several security personnel. As they neared the object in the forest communication was lost and one of the security personnel refused to approach the now visible craft. Prior to this the local police had received report that the livestock from nearby farm were acting peculiar.

Lieutenant-Colonel Halt also walked around the craft touching it and recorded what he had observed on tape along with taking several photographs. Investigator Nick Pope of the Ministry of Defense for Great Britain UFO project reviewed the tape and stated listening to what was occurring. He could hear the emotional fear and surprise in the voices of those present at the scene.

Lieutenant-Colonel Halt upon returning to the base turned all the evidence over to the Air Force. He was later told that the photographs that he had taken did not develop. A team of radiation personnel were dispatched to the area the next morning and found that the reading where the craft had landed *was 10 times above normal readings*. During the evening of December 26, 1980, a resident in the village Sudborne, which lies 6 miles from Rendlesham Forest, reported seeing a similar object hovering above his garden. Later that night the lights again were seen by security personnel in the forest.

On December 29, 1980 Staff Sergeant Penniston was called into a meeting with U.S. government agents. Some writers speculated that these agents were from the DIA (Defense Intelligence Agency) or the NSA (National Security Agency) or the CIA (Central Intelligence Agency), however; the investigative branch of the Air Force known as the OSI (Office of Special Investigation) would have had jurisdiction over this incident.

Staff Sergeant Penniston wrote out a 4-page statement of the event that transpired, which he dated and signed. During this debriefing, he was given a cover story and told that this matter was considered top secret. Other security personnel that had been involved were made to sign affidavits that this was a non-event and that nothing had occurred. Afterwards these security personnel were transferred to other bases or immediately discharged from the military.

**NOTE**: *This researcher was present at another UFO incident where the security personnel involved were held for 3 days by the OSI (Office of Special Investigation - the Air Force Investigation Unit ) and made to sign a similar affidavit that nothing had occurred under threat to their military status. Then afterwards, they were transferred to different bases so that they could not verify each other's story. (This is not an unusual occurrence, but rather standard operating procedure in the military.)*

The Ministry of Defense (MoD) for Great Britain stated that this occurrence was no real threat to national security. However, Admiral Lord Hill Norton former Chief of Defense for the United Kingdom stated, "That a Colonel of an American Air Force Base in Suffolk and his merry men are hallucinating, and for an American Air Force Nuclear Base this is extremely dangerous, or what they said happened did happen, and in either circumstances there can be only one answer, and that is that it was of extreme defense interest."

**Skeptics Theory:** A meteor was seen over southern England at that time and the light was seen in the forest causing a mistaken identity.

**Rebuttal:** The meteor was overhead, which does not account for a craft hovering several feet above a forest floor that was touched by several military personnel.

**Skeptics Theory:** Radiation reading was low according to the National Radiological Protection Board.

**Rebuttal:** *Radiation readings were said to be approximately 10 times above normal ground radiation readings per US Radiation Team.* This is hardly considered normal.

**Skeptics Theory:** Helicopters from the 67th Aerospace Rescue and Recovery Squadron based at Woodbridge RAF may have pickup a Russian Satellite known as Cosmos 749, which was widely seen at 2100 hours on December 25, 1980 and may have been seen as it was being transported back to base or the RAF may have been practicing rescue maneuvers with an old Apollo capsule.

**Rebuttal:** This researcher has co-piloted, been dropped off and picked up by helicopters while with the U.S. government and the sound of a helicopter can be heard from some distance away. Thus this security patrol would have recognized the sound of a helicopter. If a helicopter was transporting an object suspended below it, the pilot would not have allowed the object that it was transporting to enter the forest as it would have snagged on the trees and possibly causing the helicopter to crash.

**Question:** *Why would the RAF be practicing a rescue mission on Christmas Day?*

**Skeptics Theory:** An airman was driving a jeep around flashing the lights off and on at the forest.

**Rebuttal:** This researcher was present when two security personnel were brought before a military tribunal because their preliminary quiz grades for the Sergeants Exam which were below standards compared to others that were given these quizzes. The Squadron Commander allowed these two airmen to take the Sergeant's examine under the following conditions: if they failed, they would remain at their present grade level for their duration in the military and their time in service would be serving in a cleanup detail. However, if they passed the Sergeant's examine they would be promoted to Sergeant within 3 months.

Both of these two airmen passed this examine with a 90 grade point average and highest in the squadron. An investigation was then initiated and it was found that the trainers of the other personnel taking this examine, who were Sergeant in rank, had given their trainees the answers to the quizzes. The trainer of these two airmen did not give them the answers, thus making them study harder to pass this examine. These two airmen were promoted to Sergeant within three months of passing this examine, however the Master Sergeant who had called for the tribunal and all the other Sergeants that had given their trainees the answers to the quizzes were court martial and reduced a rank in grade. If this is the outcome of cheating on a mere examine, what do you think the Air Force would do to an airman pulling a prank that caused an entire nuclear base to go on alert? He would be subject to court martial and would be facing a stiff prison penalty. Therefore it is highly unlikely that any airman would pull such a prank.

**Skeptics Theory:** Oxford Ness Lighthouse has an intensity of 5 million candelas and pulses every 5 seconds, which skeptics claim that this was what security personnel were seeing in the forest.

**Rebuttal:** This Oxford Ness Lighthouse beacon *was designed and built with a protective steel plate and cover on the side facing inland since it was first built…and it has never been removed!* This was to prevent the beam from disturbing the local inhabitants. Thus this beam only spans out to sea and the only way the light can be seen is on a reflecting cloud.. Later investigations were performed without the shield and the light from the lighthouse could not be seen on or around the air base.

**Question:** *What of Lieutenant-Colonel Charles Halt and Staff Sergeant James Penniston who both claimed to have touched the craft? How could they have touched light?*

Colonel Charles Halt claims that the United States and British governments have covered up the truth about what occurred on those three nights. Colonel Charles Halt and Staff Sergeant James Penniston spoke at the National Press club in Washington D.C. on September 27, 2010 along with six other former U.S. Air Force Personnel who testified that U.S. Nuclear bases are being compromised by UFOs.

**Speculation:** This researcher has been present at nuclear facilities where UFOs were sighted and there always appears to be electrical malfunctions during these events. With regards to this incident there is one question that stands out.

**Question:** *How many photographs were taken and why were none of them able to be developed as the Air Force contends?*

**Answer:** Security personnel are trained to recognize abnormal and unusual events and record them as accurately as possible. In this event it appears that the security team performed as trained.

**Question:** *Why would a Lieutenant-Colonel and Deputy Commander submit a falsified report knowing this would subjugate him to a court martial?*

It is this researcher's opinion that after reviewing numerous material and films that this event did occur as stated by the military personnel involved.

## *Fact: The Official Lieutenant-Colonel Halt Report*

To the reader, filing a falsified report made to the United States military will buy the signatory about 20 years in prison, a $10,000 fine, loss of security clearance, loss of all pay and benefits along with bestowing upon them a dishonorable discharge. Pretty much it ruins the rest of your life.

With that said, Lieutenant-Colonel Halt filed an official report on what he found and believed to be the absolute truth on the Rendlesham Forest / Bentwaters Sighting. That same report can be read beneath these words. Lieutenant-Colonel Halt had everything to lose by doing so...and nothing to gain.

*Tell me, what are the debunkers risking when they regurgitate their preposterous and ludicrous explanations to this event? Just their mere reputations? In other words, they have nothing to lose, but everything to gain from their 15 minutes of fame on the news.*

**Following is a copy of Lieutenant-Colonel Halt's actual report.**

**DEPARTMENT OF THE AIR FORCE**
HEADQUARTERS 81ST COMBAT SUPPORT GROUP (USAFE)
APO NEW YORK 09755

REPLY TO
ATTN OF: CD

SUBJECT: Unexplained Lights

13 Jan 81

TO: RAF/CC

1. Early in the morning of 27 Dec 80 (approximately 0300L), two USAF security police patrolmen saw unusual lights outside the back gate at RAF Woodbridge. Thinking an aircraft might have crashed or been forced down, they called for permission to go outside the gate to investigate. The on-duty flight chief responded and allowed three patrolmen to proceed on foot. The individuals reported seeing a strange glowing object in the forest. The object was described as being metallic in appearance and triangular in shape, approximately two to three meters across the base and approximately two meters high. It illuminated the entire forest with a white light. The object itself had a pulsing red light on top and a bank(s) of blue lights underneath. The object was hovering or on legs. As the patrolmen approached the object, it maneuvered through the trees and disappeared. At this time the animals on a nearby farm went into a frenzy. The object was briefly sighted approximately an hour later near the back gate.

2. The next day, three depressions 1 1/2" deep and 7" in diameter were found where the object had been sighted on the ground. The following night (29 Dec 80) the area was checked for radiation. Beta/gamma readings of 0.1 milliroentgens were recorded with peak readings in the three depressions and near the center of the triangle formed by the depressions. A nearby tree had moderate (.05-.07) readings on the side of the tree toward the depressions.

3. Later in the night a red sun-like light was seen through the trees. It moved about and pulsed. At one point it appeared to throw off glowing particles and then broke into five separate white objects and then disappeared. Immediately thereafter, three star-like objects were noticed in the sky, two objects to the north and one to the south, all of which were about 10° off the horizon. The objects moved rapidly in sharp angular movements and displayed red, green and blue lights. The objects to the north appeared to be elliptical through an 8-12 power lens. They then turned to full circles. The objects to the north remained in the sky for an hour or more. The object to the south was visible for two or three hours and beamed down a stream of light from time to time. Numerous individuals, including the undersigned, witnessed the activities in paragraphs 2 and 3.

CHARLES I. HALT, Lt Col, USAF
Deputy Base Commander

## *Fact: Cash – Landrum Incident; December 29, 1980*

**Interviewee: Betty Cash and Vicky Landrum**

The story started out on the evening of December 29, 1980, with 51-year old Betty Cash driving down an isolated two-lane road in dense woods 35 miles from Houston, Texas. Along for the ride was her friend, Vickie Landrum, 57 years old, and Vickie's 7-year old grandson, Colby.

At about 9.00 p.m., the three saw a light in the distance, which in a few short minutes later became a glowing object that slowly crossed the tops of the tall pine trees. The area that they were in was surrounded by densely populated forest of pine and oak trees with occasional swamps and small lakes.

As they proceeded along their way, their initial thought was that the object was an airplane or helicopter from one of the airfields not too distant from their location. Suddenly, ahead of them loomed an immense diamond-shaped craft, which was hovering above the road and ahead of them. As the object hovered, it shot down a stream of reddish-orange flames in regular and quick intervals. Vickie would later describe it as being "like a diamond of fire."

They then got out of their car to get a better look at the craft. When they did an odd wave of heat engulfed them. As they stood watching the craft, suddenly the sky became full of helicopters. Betty said, "They seemed to rush in from all directions...it seemed like they were trying to encircle the thing." She assumed that they were from Tomball Airfield, northwest of Houston, or Ellington Air Force Base, south of Houston. The helicopters then escorted the craft from the area.

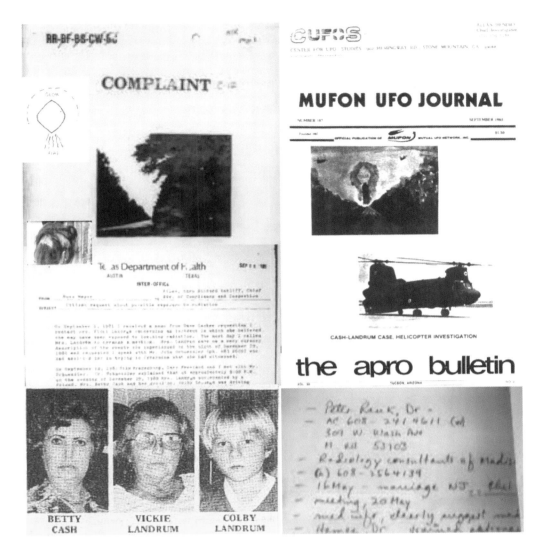

**NOTE:** In fact, the description matched the "black" CH-47 "Chinook Helicopters of the 160th Special Operations Aviation Regiment (160th SOAR), which was stationed at Fort Hood at that very time. This unit requires *at least* a "Secret" security clearance, and is used for special black-ops operations. And by a fluke meeting at an air show some time after the sighting, the incident with the UFO was later confirmed by an army pilot (in casual conversation) who actual flew that very mission! (This came to light from an investigation conducted by the UFO Hunters in 2008.)

**NOTE:** The 160th Special Operations Aviation Regiment (160th SOAR) (a.k.a. "*Night Stalkers*") only fly special operation missions regarding both secret military and para-military (CIA) missions.

After a quick trip home, all three became extremely sick within a few hours. Betty's head and neck blistered, and soon her eyes were swollen shut. She was also stricken with terrible nausea. By the next morning, she was almost in a coma. Vickie and Colby suffered very similar symptoms, but not as severe as Betty's.

After a couple of miserable days of being cared for at Vickie's home, Betty checked into a hospital where she was treated as a burn victim, remaining for 15 days. Her hair began to fall out, and her eyes swelled so badly she could not see for about a week. After some investigation, doctors concluded that the three had been exposed to high dosage of radiation exposure and poisoning. However over the years, both women had to be repeatedly admitted to a hospital for treatment.

**NOTE**: Vickie had lost about 30 % of her hair, and had large bald patches on her head. When it grew back it was of a different texture. Colby lost only a small patch of hair on the crown of his head, this too grew back in time.

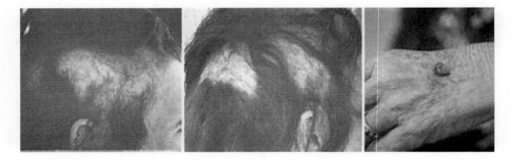

However, Betty's injuries were even worse. She experienced a severe sun burn-like condition and developed large water blisters, some as large as golf balls, over her face head and neck. One of these covered her right eyelid and extended across her right temple. She also developed a long term aversion to warm water, sunshine or other heat sources. In the year following the encounter she has spent five periods in hospital, two of those in intensive care. She lost over half of the hair on her head and has also had skin eruptions, many as big as a large coin, which left permanent scars.

**NOTE:** Most of the medical files on both women with regards to their strange illness— have mysteriously disappeared or will not be released. The women were quarantined at several times during different hospital stays with even family members denied access to visit or see them.

But Betty Cash, Vickie and Colby Landrum weren't the only witness to this. A man living in Crosby, directly under the flight path, reported seeing a large number of heavy military helicopters flying overhead as well.

Oilfield worker Jerry McDonald was in his back garden in Dayton. Although he did not see the fleet of helicopters, he did witness the huge UFO fly directly over his head. At first he thought it was the Goodyear blimp, but quickly realized it was something entirely different. *"It was kind of diamond shaped and had two twin torches that were shooting brilliant blue flames out the back"*, he said. As it passed about 150 feet above him he noted that it had two bright lights on the front of it *"like headlights"* and a red light in the center.

The Betty and Vickie eventually sued the U.S. Government for medical damages. The suit was dismissed, however, by a U.S. District Court judge in 1986 on the grounds that U.S. government involvement had not been demonstrated. This was partially due to the Department of the Army Inspector General denying any military involvement in the case, and the court disallowed any compensation for the three victims of the Piney Woods affair. Both Betty Cash and Vickie Landrum died some years later of cancer.

**NOTE**: In fact, a History channel episode of UFO Hunters ("Alien Fallout", 2008) focused on this section of roadway and even took a core sample which showed evidence of the pavement being entirely replaced. What is more baffling, is that *there are absolutely no county records attesting to this section of roadway being worked on or repaved regardless of the numerous witnesses who saw it*. This counter diction points to hard evidence of an event happening here that the military wanted to cover up completely.

## *Fact: Released Documents from a 2010 AATIP Report*
### (Supporting Cash – Landrum Incident Claim)

From filed a Freedom of Information Act request in 2017 by *The Sun*, a 1,500 page report authored by the now defunct Advanced Aviation Threat Identification Program stated,

"Sufficient incidents/accidents have been accurately reported, and medical data acquired, as to support a hypothesis that some advanced systems are already deployed, and opaque to full US understandings."

Furthermore, "The report – titled Anomalous Acute And Subacute Field Effects on Human and Biological Tissues – investigates injuries to 'human observers by anomalous advanced aerospace systems," *The Sun* reported, adding "that the report also stated that humans had been injured from "exposures to anomalous vehicles, especially airborne and when in close proximity. The report noted that often these injuries are related to electromagnetic radiation – and links them to 'energy related propulsion systems."

The report added, "Classified information exists that is highly pertinent to the subject of this study and only a small part of the classified literature has been released."

***This report confirms both United States and Alien "anomalous vehicles" do indeed exist!***

## *Rumor: 1982; Baltic Sea Area, Lake Baikal*

An underwater UFO also called a USO (Unidentified Submersible Object) was photo by a Russian submarine thru its periscope. At 150 feet the Russian captain sent seven divers out to investigate the object. When the divers got close to the USO, they saw three aliens in silver suits with helmets on, but no breathing apparatus utilized. The Russian divers decided to capture one of the aliens. However, when they approached the being with a net, a force hit them propelling them to the surface. All seven divers severely injured because they were unable to decompress. Three of the Russian divers died from these sustained injuries. However, the ones that did survive stated that the aliens were about 9 feet tall with webbed hands and feet which may have been the design of the suits that they were wearing. A description of the aliens might fall into the category of lizard like.

## *Fact: JAL Flight 1628; November 17, 1987*

Japanese Airlines Cargo Flight 1628 was traveling from Paris, France to Tokyo, Japan and while crossing the United States / Canadian border encountered a giant Unidentified Flying Object. The pilot Captain, Kensu Taraougi (with 30 years experience) radioed Air

Traffic Control in Anchorage, Alaska that he was observing white and yellow flashing lights about 8 miles in front and off to the right of his 747 aircraft. He stated that it appeared to be flying at the same altitude of his aircraft, which was 35,000 feet. Air Traffic Control at this time acknowledged that they had an unidentified blip on their radar screen. They also stated that there were no other scheduled flights in the area.

The pilot closed in on the on the unknown craft and realized that this unknown oblong flying vehicle was 4 times larger than his 747. The pilot of JAL Flight 1628 than radioed Air Traffic Control asking if he could descend to 31,000 feet and was given permission to do so. However the UFO followed the 747. The pilot than requested to make a 360 degree turn, which Air Traffic Control also granted. However upon completion of this maneuver, the UFO suddenly disappeared. It was then picked up by military radar behind flight 1628 to their left. This meant that the unknown vehicle moved to 7 miles behind Flight 1628 in a 6-second time frame thus giving it an approximate speed of 3,600 mile per hour.

This whole incident lasted 30 minutes. John Calahan, Flight Line Manager for Air Traffic Control, reported this encounter to the United States government, federal agents were then sent to Air Traffic Control in Anchorage, Alaska. These agents confiscated all the data concerning this incident and told the personnel at Anchorage, Alaska that this sighting never occurred and that these agents were never there. However, John Calahan the Flight Line Manager for Air Traffic Control made copies of this event and was willing to present this to Congress for an investigation.

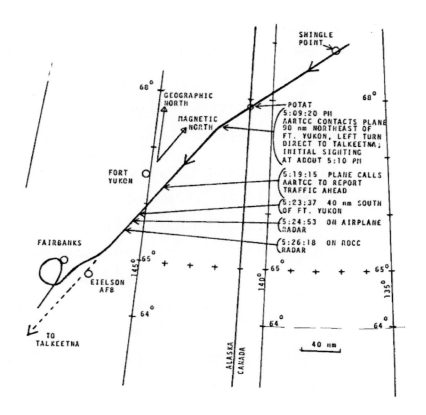

**JAL Flight 1628 Course changes during UFO encounter**

## UFO Sighting Confirmed By FAA, Air Force Radar

*Japanese Crew Tells of Encounter Over Alaska*

By Jeff Berliner
United Press International

ANCHORAGE, Jan. 1—A veteran pilot whose UFO sighting was confirmed on radar screens Tuesday said the mysterious object was so enormous that it dwarfed his Japan Airlines cargo plane.

Capt. Kenju Terauchi, the pilot, also said he saw two other small unidentified objects—smaller than his cargo carrier—that did not appear on radar.

Terauchi, his copilot and flight engineer told Federal Aviation Administration investigators that they saw the lights of an unidentified object on the evening of Nov. 17.

"They were flying parallel and then suddenly approached very close," said Terauchi, 47, who requested and received FAA permission to take whatever action was necessary to avoid the object that appeared for a time on FAA and Air Force radar and on the radar screen in the cockpit of JAL flight 1628.

The FAA confirmed on Tuesday that government radar picked up the object that Terauchi said followed his Boeing 747 cargo jet.

Terauchi, a pilot for 29 years, said he briefly glimpsed the large unknown object in silhouette. "It was a very big one—two times bigger than an aircraft carrier," he said.

Terauchi made a drawing of how he thought the objects looked. He drew a giant walnut-shaped object, with big bulges above and below a wide flattened brim.

The captain, who is stationed in Anchorage with his family, was flying the jumbo jet from Iceland to Anchorage on a Europe-to-Japan flight when the crew encountered the object in clear weather over Alaska.

Terauchi said the three unidentified objects followed his jet for 400 miles.

"It was unbelievable," he said, acknowledging that some of his colleagues have doubts about what the crew saw.

FAA investigators who questioned the crew in Anchorage concluded in a report that the crew was "normal, professional, rational, [and had] no drug or alcohol involvement." The crew's flying experience totals more than 46 years, the pilot said.

Terauchi said the crew was not frightened but wanted to avoid whatever was lit up in their flight path. "We want to escape from this."

They followed FAA directives to drop 4,000 feet and make turns—including a 360-degree turn, but Terauchi said, "They were still following us."

He said the evasive maneuvers were of no avail and the lights stayed close—once appearing in front of the cockpit.

FAA flight control reports indicate the object stayed with JAL Flight 1628 for at least 32 minutes. Terauchi said he thought it was longer. The flight controller directing the JAL plane reported the object on his radar was as close as five miles to the jet.

Terauchi said the objects moved quickly and stopped suddenly. He referred to the objects as "the two small ships and the mother ship."

Terauchi said jokingly that he thought the UFOs might have followed his chartered cargo plane because "we were carrying Beaujolais, a very famous wine made in France. Maybe they want to drink it."

KENJU TERAUCHI
... "they were still following us"

**NOTE:** Shortly after filing his report, pilot Captain, Kensu Taraougi (with 30 years experience) *was demoted to a desk job for the rest of his career.*

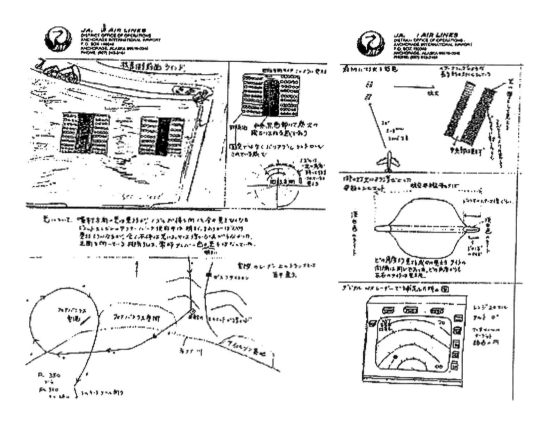

**Official Japan Airlines Report**

## *September 15, 1991; STS-48*

STS-48, Space Shuttle Discovery orbital mission transmitted a film footage of what appeared to be a UFO entering the upper atmosphere of our planet with a second floating nearby. The TV camera recording the footage was located at the back of Space Shuttle Discovery's cargo bay and was trained on the Earth's horizon while the astronauts were occupied with other tasks. Apparently two ground missiles were fired at the objects, each missile in quick sequence.

Skeptics have insisted these were merely shots of small ice particles which inevitably end up in orbit with every space shuttle. However, UFO investigators were quick to dispute this interpretation along with US scientist Richard C. Hoagland, who conclusively demonstrated the objects were actually large in size and many hundreds of kilometers away from the shuttle.

Each missile, after a brief flash, came shooting up from our planet surface. The descending UFO then made a distinct right hand turn and headed back into space as each rocket shot past. NASA claimed it was possible space junk or a meteor. A meteor has never been known to change course abruptly in an atmosphere or in space.

### *September, 1994; Zambagwa, Africa*

A UFO landed near a school's playground and was observed by 64 children along with a small alien the size of an 11 year old child. The children said that the alien had large black slanted eyes 4 times the size of ours and was dressed in a black jump suit. The alien came within 3 feet of the children, but did not speak. It may have tried to communicate with the children telepathically. When it left the children went back into the school and informed their teacher. The children were separated and told to draw what had seen. The teachers were stunned when all the drawings showed the same being and craft.

### *March 14, 1997; Phoenix, Arizona (The Phoenix Lights)*

Phoenix, Arizona has the most UFO sightings of any city in the United States. On this date one of the most renowned UFO sighting took place over Phoenix Arizona known as the Phoenix Lights. It occurred from 20:15 to 20:45 military time or 8:15 p.m. to 8:45 p.m. civilian time. It started outside of Luke Air Force Base as truck driver Bill Miron while entering the air base spied two orbs above the air base then heard the roar of two interceptors in full throttle with their after burners on go in pursuit of the orbs. The two orbs took off and the jets retuned to base.

A short time later a giant UFO said to be at least 5,000 feet from wing to wing appeared over the city of Phoenix. It was so huge that it blocked out the stars and one witnessed said it was V shaped and gun metal black. News 15 station checked with Luke Air Force Base and they denied to know anything about it. Then almost two hours later around 22:00 hours (10:00 p.m.) the Air Force started dropping flares and tried to tell the populace that the flares was what they had seen. The Governor of Arizona tried to dismiss the incident, however; when he was no longer in office he admitted to having seen the lights himself and when asked what he thought the lights were he said, "A UFO." **(Below are some of the photographs taken.)**

**NOTE:** With both authors having served in the military under various circumstances and conditions (*as well as observing and using a wide variety of military ordnance*), both can testify that these are not photographs taken of flares or any other known pyrotechnic device dropped by aircraft or dispersed by any other known means.

**FACT:** The United States military tried to hoax the people of Phoenix to believe that the earlier lights of the huge delta wing UFO seen around 8:30 p.m. was created by them at about 10:20 p.m. First understand this, the military using A-10 aircraft that night broke all safety protocols as well as established aviation law by their antics. *It was illegal then and is still illegal now for any military aircraft to fly in that vicinity due to the proximity of Sky Harbor Airport and the smaller commercial airports that surround it.*

There have been some pilots who have come forward and state that they were a part of that formation. Therefore, they should be arrested, and charged for committing the above stated crimes.

By the way, right after the event, Arizona Governor Fife Symington called the commander of Luke Air Force Base and the general of the National Guard, asking what the lights were. He was told, **"No comment."**

Second, the flares the A-10s dropped in no way matched the lighting of the UFO.

Third, the attempt of the A-10s to fly in formation dropping flares to mimic the UFO months later in June on a training exercise at the Barry Goldwater Range near Phoenix (*where they are allowed to legally to fly*) was another disaster. And in the 24 years that have passed, the military has failed to recreate what was seen and actually recorded that night by so many people.

Finally there is this, the UFO, which was seen that night over Phoenix, **was tracked on radar!** It first appeared near Area 51 traveling from the Nevada south into Arizona. It turned northwest about fifty miles south of Phoenix before disappearing. Furthermore, its timing matched the 8:30 sighting over Phoenix. There is absolutely no explanation for this. So what was really seen?

**NOTE:** During the era of the Vietnam War, an Air Force F4F fighter was training outside of Los Angeles International Airport (LAX). Due to a problem with its oxygen system and transponder, it unintentionally collided with an incoming airliner, killing all aboard the commercial aircraft and the "back-seater" (weapons officer) of the fighter jet. Only the fighter pilot safely ejected. For this reason, military aircraft do not operate in commercial flight areas unless directly ordered to do so by a higher command authority. **This is SOP (Standard Operating Procedures). Any military pilot who claims (and knowingly did) violate both military regulations as well as civilian law should be prosecuted to the fullest extent of the law for doing so.**

**NOTE:** LAX's radar at the time of the above incident was old and problematic, and was schedule for both repair and upgrade. It was part of the disaster in the making. This led into the catastrophic events of that day as well as the Air Force pilot knowingly flying an aircraft that the ground maintenance had listed for repair. **(This was an episode of the History Channel's show "Air Disasters." The fighter's wing had actually severed the cockpit of the airliner from the aircraft, which proceeded to crash into the side of a steep hill. This incident generated those regulations and laws.)**

## 2001; Tactical Air Command, Langley Air Force Base

Sergeant Karl Wolfe was as a technician working on electrical equipment temporary assigned to a secure area that reviewed classified photographs for NASA and the military. During his brief duty, he saw a lunar orbital photograph of a UFO base on the dark side of the moon. He described that there were artificial structures with a tower to the news media. A short time later he was hit by a truck while riding his bicycle and instantly killed. Sergeant Karl Wolfe held a Top Secret clearance. Is this a coincident,...*or a case of homicide?*

## 2004; Catalina Island

A white "*Tick-Tack*" shaped UFO was detected 20 miles from San Diego underwater. It was 40 feet in length and was considered in the defense zone of the aircraft carrier Nimitz. It was first picked up by radar operator Kevin Day. Two F18 Super Hornet jets were scrambled to investigate. Pilot David Fraver was first to see the object under the water close to the surface.

It was moving irritably, then it broke through the surface. The UFO hovered for a few moments. The pilots tried to pick it up on their infrared camera, but when they switched the system on it is also tied into the weapons array and that system jammed. This most likely was done by the UFO thinking that the jets were going to fire at them. It was then that pilot David Fraver switch on his regular camera The UFO then began to move against the wind with no visible sign of propulsion.

The jets were able to keep it in sight for a few moments then the UFO moved away out of the camera sights at hypersonic speed. A pilot can see fifty miles ahead from his cockpit, however; this UFO was out of sight in 2 seconds. Thus if 1 second is equal to 25 miles multiply that by 60 seconds, which puts it at a speed of 1,500 miles per second times 60 minutes equals 90,000 miles per hour. This would give it a G Force (Gravity Force) of 5,000 Gs. A human body would be turned to mush at that speed unless the vehicle had an inertia dampening system. It later came out that there was more than one incident that occurred with film footage taken. There was four in all.

The story of this incident was released to the New York Times in December of 2007 along with film footage and photographs. At that time the New York Times was also informed of a government program called AATIP (Advance Aerospace Threat Identification Program), which is used to monitor UFO activity along with any other occurrences in our atmosphere. In 2020, the Pentagon confirm that the film footage and pictures were real.

(Below is another photograph that was not presented in the Forward of this book.)

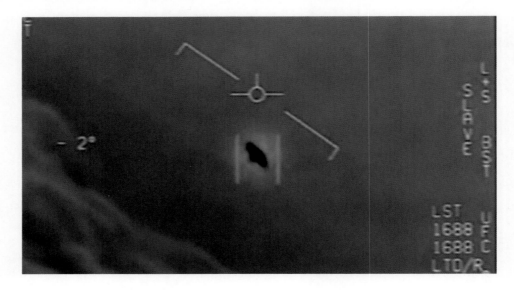

## *November 7, 2006; O'Hara International Airport at Chicago, Illinois*

A group of 12 United Airline employees consisting of supervisors, pilots, mechanics and security personnel witnessed a huge UFO over gate Charlie 17 near hanger 44. The UFO was estimated to be 530 feet in length and about 140 feet in height hovering at a 1,000 feet. It hovered for a short while before going straight up, breaking a hole in the cloud covering above. Numerous pictures were taken by passengers and employees. The FAA (Federal Aviation Association Agency) claimed that the incident it was caused by a freak weather condition and while United Airline employees were told not to discuss this incident. Clearly this presented a dangerous situation at a heavily travel Airport. **(Below are copies of the actual photographs taken of the event)**

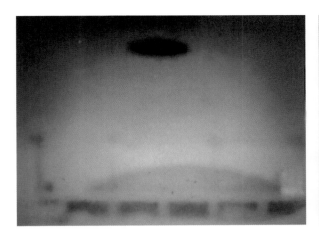 

### *July 2007, Varginha, Brazil:*

UFO Researcher Jose Gevaerd stated 2 aliens were captured as witnessed by 18 individuals in and around the above mentioned city. The first being was captured in the morning by soldiers of an army unit utilizing a net. The second alien was reported to authorities by 3 girls near a lake. The being was captured by Military Police, of which one MP died from a bacterial infection twenty days later. It was diagnosed that the bacterial infection overwhelmed and destroyed his immune system.

### *Rumor: November 20, 2013*

NASA picked up on radar and on a telescope a UFO that was estimated to be at least ¼ mile in length.

### *July 9, 2015; USS Trepang*

This American submarine was on patrol in the Arctic Ocean. A navy officer aboard by the name of John Kulka was looking through the periscope when he spied a UF/USO emerging from the water and proceeded to take a series of photographs of the object before it dove into the waters to submerge. It did not reappear for the rest of the voyage.

**(Below are copies of the UFO photographs taken through the submarine's periscope.)**

## FOX 32 News, Chicago: Published May 30, 2016

Video of a strange object in the sky over Wright-Paterson Air Force Base in Dayton, Ohio, was reported by "Fox & Friends." A couple spotted the object as they were watching the sunset from their home and recorded it.

The YouTube video (13 minutes long) appears to show an object hovering and disappearing into the clouds near the Wright-Patterson Air Force Base, which has been alleged to have UFO parts and vehicles as well as deceased extraterrestrial beings from 1947 onwards. *This particular vehicle has been sighted **at least three times flying and maneuvering over the base**.*

 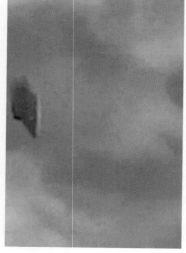

### May 19, 2017; Manatouba, Canada

A weekend prospector, Stefan Michaiak was working a claim at Falcon Lake Canada when he saw a landed disc. Thinking it was a Canadian military craft in trouble, he approach the vehicle. He found a door open and placing a gloved hand on the framed he leaned in hearing voices. He then yelled into the craft. Suddenly the voices stopped and the door started to close. The glove that was touching the craft began getting hot and started to melt just as the craft turned and a vent came into view. Hot gas then sprayed out of the vent knocking him to the ground. The gas was so hot it set his clothing on fire. The craft took off as he rolled around trying to put the fire out. Not feeling very well, he went home to his house in Winterpeg, Canada. He suffered illness for several years with a foul odor emendating from his body. He reported the incident to the local authorities.

### July 15-16, 2019; Off of Coast of San Diego, California

*This story was originally broken by Matt Adams, Duncan Phenix, George Knapp, Jeremy Corbell through* **Mystery Wire** *and Posted: May 14, 2021 before any other news outlets published the story.*

Apparently the U.S. Navy was holding military exercises when confronted with both UFOs, and USOs, one of which dove straight into the sea at 11:00 p.m. without disturbing the water in the least by this plunge, which was recorded by a sophisticated FLIR camera.

The objects, as many as 14, swarmed around the USS Omaha, a Navy "*Surveillance Ship*" on the 16th of July. The objects were recorded on the ship's S-Band Radar as well as more secret surveillance systems. The objects were either moving at speeds from 40 knots to 138 knots, or holding steady in a 40 knot head wind.

Below are images from photos and video that were taken during the two days of this encounter. The Pentagon did give a public briefing on this and released the images with information. (A submarine did investigate if the USO was either damaged or destroyed after plunging into the sea near the USS Omaha, but no debris whatsoever was found.)

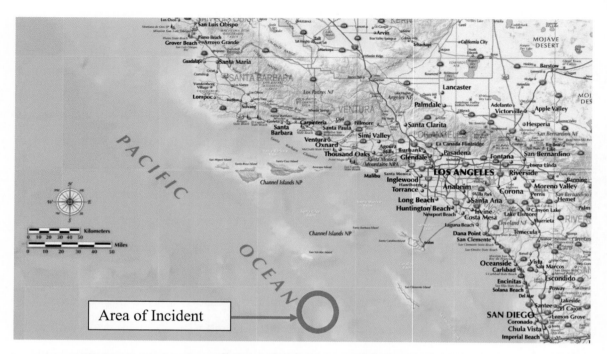

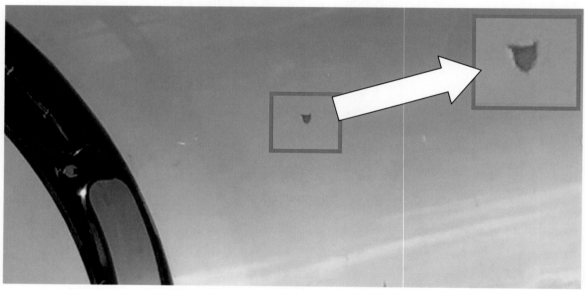

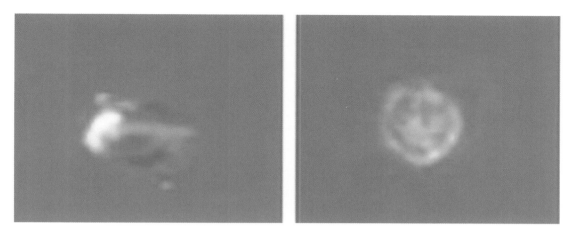

**More UFOs**

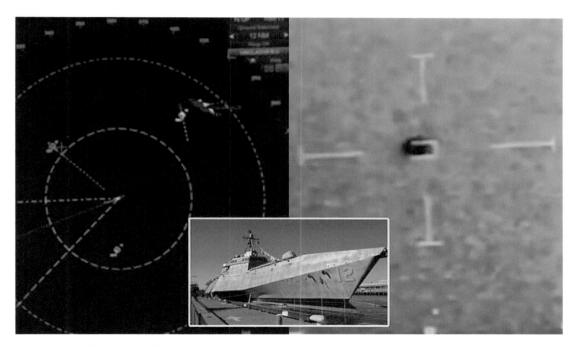

**USS Omaha with USO captured on Radar and FLIR Camera Images**

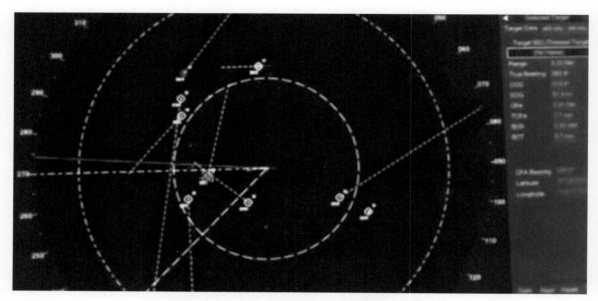

**UFO Swarm around USS Omaha on Radar**

## *Secret United States Air Force Vehicle; September 22, 2021*

A short video of the unidentified aircraft first emerged on "TikTok," a CCP (Chinese Communist Party) Social Platform. The location was tracked down by Ruben Hofs and re- tweeted by him on Twitter. The footage shows what looks like to be a stealth aircraft shape of unusual design being transported on a flatbed trailer.

The UFO-like aircraft that was seen being towed into a top-secret US military base, which accordingly was Lockheed Martin's secretive Helendale RCS Facility (radar-cross section measurement) in the Mojave Desert, California. This site is located near Lockheed Martin's Skunk Works headquarters at Plant 42 in Palmdale.

***This short video had to have been authorized and released by the United States Military itself on TikTok.*** This we believe was a specific response to the threat of Red China attacking Taiwan and/or possibly for the deliberate released of Covid-19 from the Viral Lab in Wuhan, China which was proven by varies documentation from Judicial Watch, "Five Eyes" and other official government sources.

The radar structure in the photograph appears to of an advanced redesign of Soviet Duga (over-the-horizon) Radar System, operational from July 1976 to December 1989. Two such radars were deployed, one near both Chernobyl and Chernihiv in the Ukrainian SSR (present-day Ukraine), while the other was built in eastern Siberia.

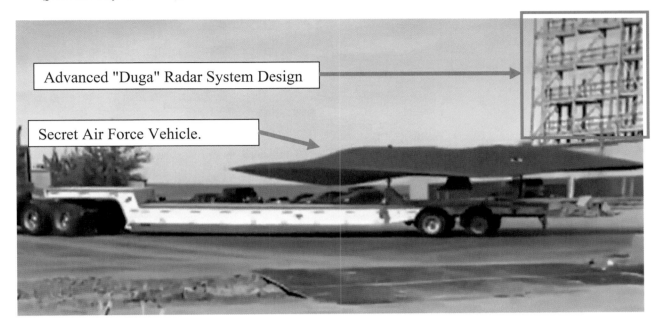

**Abandoned "Duga" Radar System in Chernobyl Exclusion Zone**

# Chapter IV: Roswell – The Cover up

## *Fact: Crash at Roswell*

**Unsubstantiated:** Dr. Vannevar Bush of the OSRD (United States Office of Scientific Research and Development) stated a UFO did crash at Roswell and alien bodies were recovered. He further stated that the government said that this was of tremendous significance and considered this to be classified higher than the H bomb.

**NOTE:** There were 70 UFO reports made in the two weeks prior to the Roswell incident.

## 21 Known Principle Individuals of Direct or Indirect Involvement:

Frank Kaufman – Counter Intelligence Corp Agent

Barney Barnett – Civil Engineer

Sheriff George Wilcox

Vern and Jean Mattais

Phyllis Mc Guire (Sheriff Daughter)

Mack Brazel (Father) Rancher

Frankie Rowe (Fireman's Daughter)

Bill Brazel (Son)

Glenn Dennis - Mortician

Major Jesse Marcel (Father)

Doctor Jesse Marcel (Son)

Sappho Henderson (Wife)

Capt. Oliver Pappy Henderson

Lieutenant Walter Haut – Press Officer

President Truman - USA

Johnny Mc Boyle – Lydia Sleppy (Reporters)

Mrs. Gardner – Top Secret Secretary

W. E. Whitmore – KGFL Radio Station

Walter Whitmore (Son)

Charles Wilhelm - Handyman

### Fact: Air Force's Initial Reaction to the Roswell Crash

Around July 3–4, 1947, an object crashed near White Sands, New Mexico. The location of the crash was the Foster Ranch outside of Roswell. Rancher Mack Brazel reported the crash to the Sheriff of that town who suggested that he inform the military. Mack Brazel notified the Army airbase at Roswell and filed his report. Major Jessie Marcel went to the town of Roswell and spoke to Mack Brazel. Upon viewing some of the wreckage that Mack Brazel had brought with him, Major Marcel then decided to take Mister Brazel back to the Army Air Base for Colonel Blanchard to make a determination on this matter. Mack Brazel after directing the military to the crash site *was placed under house detention* and Colonel Blanchard sent Major Jesse Marcel with a detail of men out to the ranch to gather the evidence.

At that time the military released a statement to the press that they had captured a flying saucer on Tuesday, July 8, 1947; which read: **"RAAF Captures Flying Saucer on Ranch in Roswell Region."** Later on July 9, 1947; there was a press release from General Ramey stating that this was all a mistake. **He said what was found was a**

**weather balloon from Project Skyhook.** (*Remember that statement.*) The story then died down until the 1978 when civilian investigators began reviewing the Roswell crash because Major Jesse Marcel (Retired) went public with his story. The government then released another story that the crash was of a top secret balloon that could monitor the Russian nuclear testing with microphones called Project Mogul. When investigators inquired about bodies found. The government had to release another story that the bodies recovered were really test dummies.

Brigadier General (Retired) Thomas DuBose admitted before his death that the balloon story was a cover up at the highest level and classified *above Top Secret.* (*There have been over 20 deathbed confessions by high ranking military personnel about Roswell's UFO crash being true. Now why would they lie when they knew they were dying, when most people try to square things with God?*)

**Fact:** Popular Science magazine in May of 1948 (on page 98) declared in a cover story article "***Are Secret Balloons the Flying Saucers?***" It was obviously not TOP SECRET in 1948 and afterwards, and its name was correctly spelled as one word "Skyhook" right from the start. **(The Cover with Entire Article are shown below in two photo montages.)**

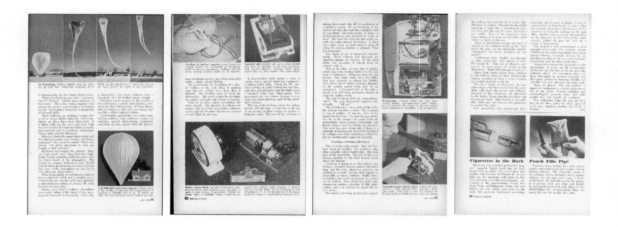

***The first official Skyhook balloon*** was launched on September 25, 1947 (almost three months ***after*** the crash at Roswell). The balloon was developed by General Mills' Aeronautical Division. Carrying an approximate 63 pounds payload of nuclear emulsion to over 100,000 feet, the balloon had a "limp" look to it. **(Below: Three Balloon Photographs and a Figure from a military manual of the Skyhook/Mogul Balloon.)**

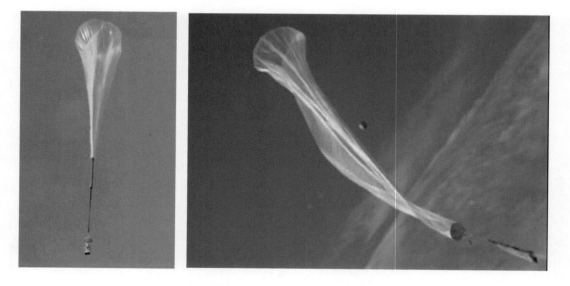

The only two MOGUL BALLOONS launched in June of 1947 from Alamorgordo, New Mexico were Flight #5 on June 5th, 1947 (which was successful) and Flight #6 on June 7th, 1947 (unsuccessfully launch).

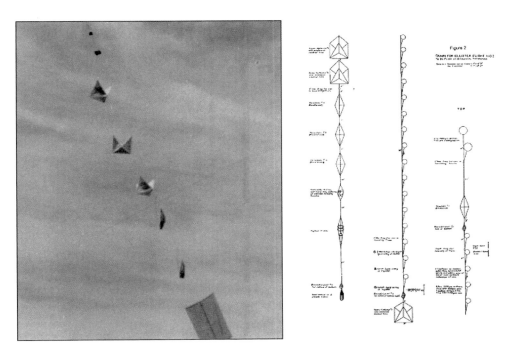

Regardless, the government has kept changing its story as to what was recovered at the Roswell crash. The first cover story Project Skyhook was a result of a previous experiment by the Navy called Project Helios. Project Helios (renamed Skyhook) only had one balloon released from Saint Cloud, Minnesota, on 25 September 1947. Skyhook payloads often consisted of cosmic-ray traps, which were stacks of photographic plates; or biological test payloads, such as seeds, mice, hamsters, and monkeys, with life support systems developed for the "passengers" to keep them from freezing or suffocating, which takes us to Project Mogul.

## *The fog of Project Mogul: A litany of secrets and misinformation*

Now at this point, things now tend get a little muddled, since the United States Air Force cannot make up its mind as to what to say about Project Mogul. Its *official* accounts of Project Mogul run the gambit of either being the forerunner of Skyhook…or its predecessor. Actually it was neither. The earliest Project Mogul balloons, however, were equipped with microphones for detection of Russian nuclear tests…and later were used to drop crash-test dummies for parachute development in 1952. And the majority of Project Mogul flights were conducted from the air base at Alamogordo, New Mexico.

On the surface, it appears that Project Mogul was the Army Air Force's (later United States Air Force's) equivalent to the Navy's Project Skyhook. According to some documentation, Project Mogul transformed into Project Skyhook, which then reverted back into Project Mogul again and was renamed Project Gopher. *Huh?* But this mumble-jumble may have been the work of some confused Department of Defense typist or clerk. God knows that wouldn't be surprising.

In either case, balloons were used. For Project Mogul, balloons consisted of an array of many small balloons tethered together, while Project Skyhook used single balloons approximately 30 meters in diameter. Some of them were quite long, stretching out over five hundred feet. Yet, they both achieved the mission of operating at high altitudes.

For Project Mogul, there are some past hand-written records that show the first successful flight a month before Roswell happened. *Flight 5* was that balloon array, launched on June 5th, 1947…and was the first successful flight of Project Mogul (meaning all the other previous flights failed to launch or crashed soon after). **(Below: ACTUAL photographic copy of Project Mogul Balloon Flight Logs with the "Balloon Flights in Question.")**

| FLIGHT NUMBER | DATE AND RELEASE TIME | LAUNCHING SITE | DESCRIPTION OF BALLOONS | CRITIQUE |
|---|---|---|---|---|
| A | 20 Nov. 1946 1438 AST | NYU, N.Y. | 2 - 350 gram meteorological | Balloon balancing load. Free lift from 350 gram meteorological balloon. Successful cutting free of lifter balloon. Balloon did not level off. |
| B | 16 Dec. 1946 1219 EST | NYU, N.Y. | 2 - 350 gram meteorological | Balloon balancing load. Free lift from 350 gram meteorological balloon. Successful cutting free of lifter balloon. Balloon did not level off. |
| 2 | 3 April 1947 1412 EST | Bethlehem Pennsylvania | 14 - 350 gram meteorological balloons. Long cosmic ray train | Failure due to poor rigging, poor launching technique. 2 lifter balloon 12 main balloons. Train rose until some balloons burst then descended rapidly. |
| 5 | 5 June 1947 1917 MST | Alamogordo New Mexico | 29 - 350 gram meteorological balloons. Long cosmic ray train | First successful flight carrying a heavy load. 3 lifter balloons, 26 main balloons. |
| 6 | 7 June 1947 0509 MST | Alamogordo New Mexico | 28 - 350 gram meteorological balloons. Long cosmic ray train | Flight unsuccessful. Altitude control damaged on launching. 4 lifter balloons, 24 main balloons. |

The *"cosmic train"* payload *Flight 5* carried would have easily been identified as such by the airmen of the 509th. For instance, one of the fragile balsa wood struts used as one of the frames of the radar targets looked more like a child's kite than a flying disk. Then there were the instruments attached to the nylon line, which were not anymore mysterious than anything found at a radio repair shop: a ballast reservoir, an electric battery, a radio transmitter, and a "sonobuoy," which looked like a mere metallic can. (*Furthermore, none of these instruments were found on the Foster Ranch, neither by the rancher Mack Brazel, nor by the military who later came to retrieve the debris.*)

**Question:** *The military has maintained that the balloon that crashed at Roswell was Flight 6, which attempted to launch on June 7th, 1947. However,* **by their own records**, *Flight Number 6 was a launch failure and immediately aborted! Since Flight 6 was aborted, how did it crash in Roswell a month later when it never left the base?*

**Answer:** It didn't, however; the wreckage of a weather balloon that General Ramey displayed (*as Major Marcel admitted prior to his death, which was switched with the actual UFO wreckage*) could have very well been that of *Flight 6*.

**Question:** *Could the wreckage have been of Flight 5, launched on June 5th, 1947 (which transmitted for days on end)? And if so, why wasn't it discovered earlier by Mack Brazel,* **who checked that very same land nearly on a daily basis***? Surely, he should have discovered the wreckage at least three to four weeks prior if it was that very same balloon. And that should have been extremely easy since the US Army officially declared* **"the debris field scattered over an area of about three quarters of a mile long and several hundred feet wide."** *(That's a little hard to miss, wouldn't you say? Any way would the debris from just a balloon take over a platoon of men days to police up?)*

**Answer:** No. Since *Flight 5* made its intended altitude, it would have been traveling at approximately over 200 miles an hour because of the jet stream. Since Roswell is roughly 100 miles northeast from the balloon's launch point at Alamogordo AFB in New Mexico, the calculated flight time would be only about thirty minutes. And since it was broadcasting for a period of days, there was no way *Flight 5* crashed at Roswell since it would have been over a thousand miles away *in 5 hours!*

If it had crash, then it would have been immediately detected by radar tracking and with the transmitter still working, the wreckage should have been recovered easily within days of June 5th. But that didn't happen! So whatever came down, came down within a few days of July 3th or 4th…*and it wasn't any weather balloon either!* Simple logic dictates a balloon couldn't have been loitering above Roswell for over four weeks in winds exceeding 200 mph. And then there is the size of the debris field, which was way too large to be the balloon array and its payload; it therefore had to have been something else…some type of large aircraft…or perhaps...a space vehicle. And since there were no reported missing aircraft, what does that leave?

**Question:** *If you believe the government that it was just a simple weather balloon crash, why would the government then classify the incident* **"above Top Secret"** *when they publicized these balloons extensively at air force shows and throughout media outlets?*

**Answer:** The capture of a real Unidentified Flying Object and its occupants would have been classified above Top Secret…and not these balloons.

**Question:** *About the bodies found at Roswell, the government stated that these were either chimpanzees or test dummies that were found. But the time-line is all wrong. How could chimpanzees or test dummies be at a crash site in 1947 when they weren't used until 1950 and 1952, respectively?*

**Answer:** Since the government did not begin to use chimpanzees until 1950 and test dummies until 1952, there is no accounting for the "*child-like*" bodies found at the crash site reported by both civilian and military personnel. We seriously doubt that they drop through a time portal back to 1947 or does the Air Force have an actual secret way of transporting chimpanzees and crash dummies exclusively back in time that they're not telling us about?

**(Below are two MILITARY photographs of the test dummies, which are 6 feet in height. They hardly match the description of 3 to 4 foot aliens seen at the crash site and as in the purported frame from a MILITARY film. Can you tell if there is any difference?)**

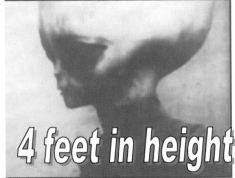

On the left is a picture of the dummies used in the high altitude experiments. On the right is a frame of film which the witnesses described at the crash site. According to the Air Force they look exactly alike!

We'd surely like to hear their explanation about this, but think the Air Force should at least get its stories straight before issuing any further ridiculous and idiotic statements on the subject. The blather of nonsense is rather demeaning for them…and insulting to the public!

## *Fact: Radar Screens*

**Interviewee: Frank Kaufmann - Army Counter Intelligence Corp Agent**
July 3, 1947 - Radar Screens at Roswell Army Airbase home of the 509th Bomb Wing picked up erratic movement of blips that seemed to pulsate then exploded and

disappearing off the radar screen. (Sounds like a dogfight, doesn't it?) Location was given as east of White Sands, New Mexico (where most of our nuclear testing took place) by a motorist traveling to Roswell on highway 285. They thought that an aircraft had crashed because they noticed a flaming ball falling to earth. Receiving the call, military police were dispatched from Roswell Army Air Field. Upon arriving the military police found a group of archeologist students and a civil engineer (Barney Barnett) at the scene of a second crash site.

## Fact: 2nd Wreckage Site – Major Crash Site

**Frank Kaufmann** (a supposed Army Counter Intelligence Corp Agent), who was dispatched to the crash site, described the scene of the incident. A triangular shaped craft was imbedded at a 30 degree angle into a hill on an otherwise flat land. The impact site was 30 miles north of the airbase. It illuminated an orange glow indicating possible radioactive contamination on top while the bottom of the craft was emanating a soft white light. The craft was 25 feet long and 12 feet wide. The left side of the craft was split open.

Military personnel in chemical suits arrived and checked for radiation readings. Then the military photographers began making a photographic record of the crash site. There were two bodies on the ground near the craft and three bodies inside the craft all about 5 feet to 5 feet 5 inches in height. Their heads and eyes were larger then our own, but the rest of their bodies were in proportion. Their skin was ash in color. Their hands showed some signs that deterioration was starting to occur and because of this the bodies were placed in black body bags. The bodies were then driven in hospital trucks to the Roswell Army Air Field. The wreckage was also taken to the airbase and brought to hanger 84.

**Question:** *If the downed craft had been an AAF (Army Air Force) experimental predecessor to USAF (United States Air Force) flying wing, how much of aircraft would have been intact after hitting the ground?*

**Answer:** Not too much of it, recall that the impact was at a 30-degree angle straight into the side of a hill. If you look at FAA and military footage of simulated controlled tests of airplane crashes, you'll see that every one of these aircraft disintegrated from the force of the impact, while the rest was consumed by explosion and fire of its fuel. Neither of which was observed at this crash site with the vehicle mainly intact. Ergo, it wasn't an AAF flying wing. Keep this in mind when reading about the properties of the debris found.

## *Rumor: Number of Bodies*

The number of bodies seems to differ in several different testimonies from two to six bodies recovered. It was also rumored that two of the beings (*which means there had to have been more alien beings*) were still alive when the military police arrived and that a MP Lieutenant beat one of the aliens to death with the butt of his carbine in order to get a black box that the alien was holding.

**Speculation:** The remaining alien, according to *Falcon* and *Condor*, two supposedly intelligence sources that work in Area 51 (credentials purportedly verified by "UFO Cover Up Live" producers, which was a television documentary that aired in November of 1987), was from the star system of Zeta 1 Reticuli. (*Why does this star system keep cropping up in so many UFO cases?*) Coincidentally, this was one of the base star systems in Marjorie Fish's star map that she comprised from Betty Hill's drawing which will be discussed later in this book.

**NOTE:** Lieutenant Governor of New Mexico (1947 to 1957) claimed to have seen the alien bodies and said that they were not human.

## *Fact: Civil Engineer (2nd Hand Information)*

### Interviewee: Vern and Jean Mattais

Barney Barnett lived in Socorco, New Mexico and worked as a civil engineer for the U.S. Soil Conservation Service. One day in 1950, while visiting his close friends Vern and Jean Mattais, he told them that he had come upon a crash site of a strange craft with dead alien bodies on July 4, 1947. He professed that shortly after, the military had arrived and closed off the area after swearing everyone there to secrecy. He then related the story summarized below:

**Summary:** Barney Barnett stated that he was on assignment directly west of Soccrco in the Magalalena region in the plains of San Aqustin, when in the distance he saw a glimmering light. Thinking that it was a downed aircraft, he went to investigate too see if he could render assistance. Upon arriving at the crash site, he came upon a large disc shaped object about 30 feet in diameter. Just then some archeology students from the university of Pennsylvania or Michigan arrived at the crash site. They too thought that it had been an airplane that had crashed. Upon taking a closer look, he spied 4 bodies on the ground and others in the crashed vehicle, which appeared to have been split open by an explosion or on impact. The bodies were human-like, but not human, with large heads and small eyes oddly spaced. They were quite small and had no hair. They wore one piece gray suits and all appeared to be male. He stated he was close enough to touch them, but did not. ***It was then that the military arrived and closed off the area swearing everyone there to secrecy explaining it was their patriotic obligation.***

**Speculation:** It appears that Barney Barnett story coincides almost perfectly with that of Counter Intelligence Corp Agent Frank Kaufman. However, these two individuals were not friends or known to each other.

**Question:** *Why would both men (strangers to each other) tell the exact same story, unless the incident really occurred?*


## *Fact: The 1st Wreckage Site – Minor Crash Site*

William Mack Brazel worked on huge ranch with eighty sections that contained 700 head of cattle and about 4,000 sheep. Because of its size he would inspect the property constantly. On July 3, 1947 there was a terrible lightening storm. Mack Brazel watched the storm from his house and noticed that the lightening seem to be striking the ground over and over again in the same spot as if something was drawing the electricity to it.

On July 4, 1947 Mack Brazel went out to inspect his property to see if any damaged had been caused by the lightening, especially the area where it had struck repeatedly the night before. When he arrived where he thought the lightening had struck, he found that over nearly a mile of the area heavily littered with wreckage and debris. It appeared as if something had blown apart with tremendous force above the land. He noted that none of the sheep would go near the debris field.

Mack Brazel then picked up some of the wreckage and took it to his closest neighbors Floyd and Loretta Proctor. His neighbors convinced Mack Brazel to report the incident to the sheriff.

On Sunday July 6, 1947, Mack Brazel went into town to show the material he had found to the sheriff and along the way he stopped at the local bar. He showed the debris to some of the local patrons before reporting the wreckage to the Chaves County Sheriff, George Wilcox. The sheriff, after hearing Mack's story, suggested that he report this incident to the military at Roswell Airbase. **Upon filing his report with Roswell Air Base Mack Brazel was then taken into custody and held on the base for three days until all the wreckage had been removed.** (*By the way, this is an illegal act under federal law at that time and in violation of the Constitution of the United States dealing with unlawful imprisonment.*) That assignment was given to Major Jesse Marcel and a detachment of military personnel.

**Interviewee: Bill Brazel (Son)**

Bill Brazel, upon hearing his father had been detained by the military, went over to his father's ranch to assist while his father was gone. During this time, he also found some of the wreckage. He said that some of it was a tinfoil like substance, but looked more like lead foil. Balling it up in his hand, he put in his pocket, then when he reached the barn he took the substance out and placed it on the workbench. Too his amazement, it began to unroll by itself, flattening out. He also found some monofilament type object similar to a fishing line. However, it was not any type of string or wire. There was also something like balsa wood with symbols on it that reminded him of the type of symbols that are on Chinese fireworks. He tried to burn it, but it would not burn. He then tried to whittle it, but it would not cut. He later told some people about this wreckage that he had kept and the word got back to the Army Air Force. They went to his house and demanded the material. **Later an F.B.I. agent called him up and threatened to put him in jail if he ever told anyone about this incident.**

**Speculation:** It appears as if the craft had fabricated of some sort of highly conductive material (*and/or a polarized electromagnetic field*) that caused the lightening to strike it repeatedly. This must have split the craft and caused the wreckage to fall on the Foster ranch the 1st wreckage site. The occupants at that point must have lost control of the vehicle and crashed at the 2nd wreckage site.

## *Fact: The 1st Wreckage Site - Continued*

**Interviewee: Major Jessie Marcel**

Major Jessie Marcel was an intelligence officer stationed at the Roswell Army Airbase with the 509th Bomb Group in July of 1947. He was familiar with every type of balloon, aircraft, and missile in the American and Soviet military inventories at that time. On July 7, 1947, the base received a phone call from the Chaves County Sheriff Office. Sheriff George Wilcox stated that rancher Max Brazel found some sort of aircraft wreckage on his property and brought some of it to town. Major Marcel then went to town and spoke with rancher Mack Brazel. After reviewing the material that Mack Brazel had brought with him, Major Marcel determined that they should go to Roswell Army Air Field and speak with Colonel Blanchard.

After the meeting, Colonel Blanchard decided that rancher Mack Brazel would show the military where the crash site was located then return to the airfield and remain on the airbase while Major Jessie Marcel took a detail of men and remove all the wreckage (which was more than ¾ of a mile in length) from that strip of land.

Upon reaching the 1st wreckage site he examined the material more closely and came to a conclusion *that it was not anything that was made on this planet!* After the material was removed from the Foster ranch, Colonel Blanchard ordered Major Marcel to escort the material to Carswell Air Field in Fort Worth, Texas for analysis. The material was loaded onto a B-29 bomber (*turned into a transport*) and upon its arrival at Carswell Air Field, the debris was stored in hanger 18 (a hanger that was known for the housing of top secret aircraft). At this point General Ramey called a press conference and told Major Marcel not to speak with the press that he (General Ramey) would do all the talking. General Ramey left a few insignificant items from the crash site on the floor mixed in with substitute material from a real weather balloon. They then held the press conference while the rest of the material was flown to Wright - Patterson Air Field in Dayton, Ohio for further analysis.

Major Jessie Marcel was interviewed after he retired from the military about the Roswell wreckage. He said that when he arrived at the Foster ranch, he found the wreckage spread about ¾ of a mile in length and about several hundred feet wide. He stated that General Ramey made up the story of the weather balloon and that the real material was from an unknown origin and not of this earth. He further stated that the material consisted of small beams about ½ inches across with a kind of pink/purplish writing in a type of hieroglyphics on it. These beams seemed like balsa wood and weighed about the same, but were flexible and very hard. The beams could not be cut or burned. There was also some type of parachute substance that was also extremely strong and could not be burned. A great deal of tin foil like substance was also spread across the area. There was one piece that was 2 feet by 1 foot in length that was extremely thin and weighed almost nothing, but could not be pierced or dented even with a 16 pound sledge hammer, yet it was flexible and wobbled when wiggled. He later stated that he did see alien bodies and drew a picture of the aliens which matched what Betty Hill described in her abduction.

**(A drawing of the hieroglyphic symbols as remembered by the late Major Jessie Marcel.)**

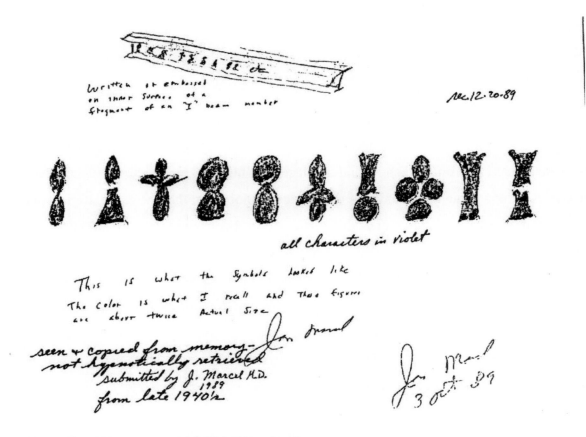

### Interviewee: Doctor Jessie Marcel Jr. (son)

Doctor Jessie Marcel Jr., Major Marcel son, was only 12 years old when the crash occurred. He recalls on the night of July 7, 1947, his father came home very late and woke him and his mother up. His father was extremely excited. His father was talking about a crash that had occurred and brought some of the material at the crash site in the house spreading it on the kitchen floor. Doctor Marcel specifically remembers the small beams with the purple hieroglyphics. His father told them that they were not to speak of this to anyone. The next morning his father loaded the material in a carryall trailer and took it back to the army airbase at Roswell.

## *Fact: The Press Release to the Public*

### Interviewee: Lieutenant Walter Haut

July 8, 1947, Lieutenant Walter Haut was called into Colonel Blanchard office and advised that a flying saucer had been captured and advised to set up a press release for this story. Lieutenant Walter Haut then went back to his desk and wrote up the press release. He was denied going to the crash site by Colonel Blanchard, thus he had to write the article without any real details. Because of the time of day it ran in the late edition of the Roswell Daily Record. On July 8, 1947, the release read as follows:

## "RAAF Captures Flying Saucer on Ranch in Roswell Region"

**"The many rumors regarding flying disc became a reality yesterday when the intelligence office of the 509th Bomb Group of the 8th Air Force, Roswell Army Air Field, was fortunate enough to gain possession of a disc through the cooperation of the local ranchers and the Sheriff's office of Chaves County, New Mexico."**

The next day the military retracted the statement saying it was a weather balloon. However, by this time the base was receiving telephone calls from all over the world. Lieutenant Haut was reprimanded for releasing the story that Colonel Blanchard had ordered.

**NOTE:** On 12/26/2002 Lieutenant Walter Haut signed an affidavit attesting to the UFO crash at Roswell, which was to be released upon his death stating that he did see an egg shaped craft with small alien bodies in hanger 84 at Roswell. He stated that these bodies had larger heads and almond shaped eyes. This affidavit was not to be released until after his death for fear of government reprisals. In 2007, this affidavit was released to the public. **The affidavit is shown in its entirety below**.

DATE: December 26, 2002
WITNESS: Chris Xxxxx
NOTARY: Beverlee Morgan

(1) My name is Walter G. Haut.

(2) I was born on June 2, 1922.

(3) My address is 1405 W. 7th Street, Roswell, NM 88203

(4) I am retired.

(5) In July, 1947, I was stationed at the Roswell Army Air Base in Roswell, New Mexico, serving as the base Public Information Officer. I had spent the 4th of July weekend (Saturday, the 5th, and Sunday, the 6th) at my private residence about 10 miles north of the base, which was located south of town.

(6) I was aware that someone had reported the remains of a downed vehicle by midmorning after my return to duty at the base on Monday, July 7. I was aware that Major Jesse A. Marcel, head of intelligence, was sent by the base commander, Col. William Blanchard, to investigate.

(7) By late in the afternoon that same day, I would learn that additional civilian reports came in regarding a second site just north of Roswell. I would spend the better part of the day attending to my regular duties hearing little if anything more.

(8) On Tuesday morning, July 8, I would attend the regularly scheduled staff meeting at 7:30 a.m. Besides Blanchard, Marcel; CIC Capt. Sheridan Cavitt; Col. James I. Hopkins, the operations officer; Major Patrick Saunders, the base adjutant; Major Isadore Brown, the personnel officer; Lt. Col. Ulysses S. Nero, the supply officer; and from Carswell AAF in Fort Worth, Texas, Blanchard's boss, Brig. Gen. Roger Ramey and his chief of staff, Col. Thomas J. DuBose were also in attendance. The main topic of discussion was reported by Marcel and Cavitt regarding an extensive debris field in Lincoln County approx. 75 miles NW of Roswell. A preliminary briefing was provided by Blanchard about the second site approx. 40 miles north of town. Samples of wreckage were passed around the table. It was unlike any material I had or have ever seen in my life. Pieces, which resembled metal foil, paper thin yet extremely strong, and pieces with unusual markings along their length were handled from man to man, each voicing their opinion. No one was able to identify the crash debris.

(9) One of the main concerns discussed at the meeting was whether we should go public or not with the discovery. Gen. Ramey proposed a plan, which I believe originated with his bosses at the Pentagon. Attention needed to be diverted from the more important site north of town by acknowledging the other location. Too many civilians were already involved and the press already was informed. I was not completely informed how this would be accomplished.

(10) At approximately 9:30 a.m. Col. Blanchard phoned my office and dictated the press release of having in our possession a flying disc, coming from a ranch northwest of Roswell, and Marcel flying the material to higher headquarters. I was to deliver the news release to

radio stations KGFL and KSWS, and newspapers the *Daily Record* and the *Morning Dispatch*.

(11) By the time the news had hit the wire services, my office was inundated with phone calls from around the world. Messages stacked up on my desk, and rather than deal with the media concern, Col. Blanchard suggested that I go home and "hide out."

(12) Before leaving the base, Col. Blanchard took me personally to Building 84, a B-29 hangar located on the east side of the tarmac. Upon first approaching the building, I observed that it was under heavy guard both outside and inside. Once inside, I was permitted from a safe distance to first observe the object just recovered north of town. It was approx. 12 to 15 feet in length, not quite as wide, about 6 feet high, and more of an egg shape. Lighting was poor, but its surface did appear metallic. No windows, portholes, wings, tail section, or landing gear were visible.

(13) Also from a distance, I was able to see a couple of bodies under a canvas tarpaulin. Only the heads extended beyond the covering, and I was not able to make out any features. The heads did appear larger than normal and the contour of the canvas over the bodies suggested the size of a 10-yearold child. At a later date in Blanchard's office, he would extend his arm about 4 feet above the floor to indicate the height.

(14) I was informed of a temporary morgue set up to accommodate the recovered bodies.

(15) I was informed that the wreckage was not "hot" [radioactive].

(16) Upon his return from Fort Worth, Major Marcel described to me taking pieces of the wreckage to Gen. Ramey's office and after returning from a map room, finding the remains of a weather balloon and radar kite substituted while he was out of the room. Marcel was very upset over this situation. We would not discuss it again.

(17) I would be allowed to make at least one visit to one of the recovery sites during the military cleanup. I would return to the base with some of the wreckage which I would display in my office.

(18) I was aware two separate teams would return to each site months later for periodic searches for any remaining evidence.

(19) I am convinced that what I personally observed was some type of craft and its crew from outer space.

(20) I have not been paid nor given anything of value to make this statement, and it is the truth to the best of my recollection.

THIS STATEMENT IS TO REMAIN SEALED AND SECURED UNTIL THE TIME OF MY DEATH, AT WHICH TIME MY SURVIVING FAMILY WILL DETERMINE ITS DISPOSITION.

Signed: Walter G. Haut

## *Fact: Civilian News Report*

### Interviewee: Lydia Sleppy

Johnny Mc Boyle was a reporter for KOAT radio station in Albuquerque, New Mexico and co-owner of the sister station KSWS in Roswell, New Mexico. On July 7, 1947; he called Lydia Sleppy the administration assistant and teletype operator at the KOAT radio station. Excitedly, he told her he had scooped a story on a flying saucer crash. He wanted her to get this to ABC wire immediately. He stated that a flying saucer had crashed on a ranch in Roswell and the military was closing off the whole area. He said that someone stated that little men were on the craft. Lydia started to send the message by teletype when her transmission was cut off by another transmission stating "**Do Not Transmit this Message – Stop Communication Immediately – The F.B. I.**" Worried about losing their FCC (Federal Communication Commission) license, she ceased the transmission.

## *Fact: W.E. Whitmore, Owner of KGFL Radio*
**Interviewee: Walter Whitmore**

Disc Jockey Frank Joyce was first to get the story of the Roswell UFO crash from Walter Haut. W.E. Whitmore was the owner of KGFL radio in Roswell, New Mexico interviewed Mack Brazel. His son Walter Whitmore remembers that his father had Mack Brazel at the Whitmore house in order to get an exclusive interview while the military were looking for Mack Brazel all over town. His father recorded the story and tried to get it on the mutual wire service, but was stopped by a man named Slowie from the FCC (Federal Communication Commission) in Washington D.C. *His father was told that this was a national security matter and if he wanted to retain his FCC license he would cease transmission immediately.* Walter Whitmore said that his father never actually saw the crash site, but he did see some of the wreckage. His father described the tough foil like substance and the small beams with the hieroglyphics printed on them. His father did go to the crash site, but was stopped by the military police. *They told him to turn around that this was a Top Secret area.*

**Question:** *If this was just a weather balloon why was it considered **a Top Secret Area?***

**Speculation:** A second question then comes to mind, why would the FCC (Federal Communication Commission) get involved and threaten to take away a radio station's license unless the government was trying to cover up something more important than a weather balloon.

**Question:** *Since when is a weather balloon considered a matter of National Security? At this point in time, none of the wartime regulations were in effect, **none** of the National*

*Security Acts had been passed, and therefore, these government entities were breaking the law! In other words, something more than just a weather balloon crashed.*

Radio reporter Jug Roberts stated that weather balloons were a common thing to find and everyone knew what they looked like.

## *Fact: Chaves County Sheriff's Office*

### Interviewee: Phyllis McQuire

Sheriff Wilcox and his family lived directly above his office and the jail. News about the crash had spread all over the town. On July 8, 1947, his daughter Phyllis McQuire (married name) after reading the evening newspaper asked about the crash. Sheriff Wilcox told her about the objects that Mack Brazel had brought in. He described the tin foil like substance and how after putting it in any form then placing it down the metal would unfold before his eyes and flatten out. The paper called it a flying saucer and her father confirmed that it was unusual. He told her that people from all over the world were calling him. He said that the military had stopped by and picked up the stuff that Mack Brazel had left at the jail. According to Phyllis McQuire an Army Air Force Officer also stopped by and had some harsh words for her father on more than one occasion. ***Eventually due to the military constant harassment Sheriff Wilcox resigned.***

After her father had passed away in the early 70's, her mother told her more of what happened that day. She said that her father stated while at the crash site he viewed three aliens bodies that were 4 to 5 feet in height. One of the aliens was still alive, but eventually died. They were small in size with large heads. She said that her father felt sorry for them because the army air force was treating them as enemies. The military told her father that they are not suppose to talk about this with anyone...*or the government would kill the whole family!* **She said that other people in the town were threatened** *in the same way too!*

**Question:** *Alternatively, you have to ask yourself the following questions, if it was a balloon with dummies, then why would the United States government threaten to kill all the people that claim to have seen these so-called dummies?*

**Answer:** Apparently this was something more than just an ordinary weather balloon and dummies, where the government felt the need to violate federal law, oaths of office, and the Constitution itself. It had to something so earth shattering that the measures taken had to be...*illegal and extreme!*

Some of the claims also viewed aliens alive and walking around. If these were dummies how could dummies walk around? Several testimonies about observing aliens alive and dead were given by individuals on their death beds, which in a court of law and giving testimony in regards to observing a murder would be considered the truth and convict the suspect in that trial. How can all the witnesses be mistaken in what they had seen?

## *Fact: Fire Chief's Daughter*

### Interviewee: Frankie Rowe

Frankie Rowe (Fire Chief's Daughter) states that her family wasn't afraid of the space craft that crashed or its occupants, *they were afraid of the United States Government.* In July of 1947 she was 12 years old and it was dinnertime, her father had just returned from his shift extremely excited. He said that morning they received a call at the fire station that there was a crash 30 miles north near Black Water Draw. He said that they thought a plane had crashed and thought that it would start a fire, so they responded immediately.

When they got there, they starred in amazement at a flying saucer. Two of the crew outside the craft were dead, but a third was walking around outside the craft. He said that they were little people and the one that was alive talked to them in their heads (telepathy) without saying anything, but they all heard the same thing. It said that it was sad over the loss of its comrades and that it wasn't here to hurt us. After that the military arrived. They were angry that the fire department was there and chased the firemen away sealing off the area. Her father said that the aliens were about 4 feet tall and looked child like with some insect like features. They were a pale pinkish gray in color.

A few days later Frankie Rowe was at the fire station when a police officer that her father knew walked in the back door. The police officer said he had something that he took from the crash site and pulled out a wadded up piece of foil. Putting it on the table it gradually unfolded by itself and was smooth as glass. He couldn't cut it or burn it. The firemen there also tried to cut and burn it without any results. They left it on the table and Frankie began to play with it.

A few days later a military car came down the road and stopped at their house. A military police officer got out along with another man in a fancy dressed uniform. His coat was a drab olive green and his pants were light pinkish gray. He wore a belt over his shoulder and around his waist. He had white spats (called *leggings*) over his black boots and sunglasses on that he never took off. He wanted to talk to Frankie Rowe and told her to sit in the dining room while he stood over her. He told her that she didn't see anything and being truthful she said that she did handle the foil. He took out his baton and began slapping in his hand as he spoke. Again he told her angrily that she didn't see or handle anything. He then said that...*if she talked about this to anyone that the government would take her and her family out into the desert and kill them!* Crying, Frankie Rowe promised not to talk about it. Frankie Rowe only began to tell her story in the 90's when she met with reporter Kevin Randle.

## *Fact: The Roswell Mortician*

### Interviewee: Glenn Dennis

Glenn Dennis was the mortician in the town of Roswell, New Mexico. He became involved in this incident when he received a telephone call from Roswell Army Air Field. Their mortuary officer wanted to know if he had any children's caskets about 4 feet long that could be hermetically sealed. Glenn told the officer that he had two. The officer wanted to know if Glenn could get more. Glenn told him that he could, but it would not be available until that evening. Glenn asked why and the officer said that this was for a discussion about an exercise. Forty-five minutes later that same officer called back and wanted to know about embalming fluid and what effect it would have on a body. He also inquired about bodies that were exposed to the elements and predators. He also asked about how to transport bodies without embalming. Glenn told him to try packing them in dry ice and where they could get it.

A short time later Glenn received a call that he was needed to transport an injured airman to the base infirmary. Glenn picked up the airman and drove him to the base. They went into the infirmary passing two open doors with wreckage on the floor with some canoe shaped metal with hieroglyphics written on them. Going further down the hall they ran into a commotion with several officers giving commands. It was then that one of the officers confronted Glenn when he inquired that this appeared to be a crash. The officer started poking Glenn and threatening him saying...*if he talked about what he just saw that dogs would be picking his bones out of the desert!*

**NOTE:** With all the reported death threats given individually by numerous witnesses, it is clearly more than a pattern, but a truthful... *and disturbing recounting of events...of crimes of blackmail and coercion committed by military personnel in violation of military code of conduct and the Constitution of the United States!*

Then two military police came over and escorted Glenn back down the hallway where he ran into a nurse he knew exiting a room. She was sobbing and gasping for air with a towel covering her nose and mouth. She told him to get out of there as fast as he could. Several other medical personnel followed her out of the room also gasping as if they were about to throw up. Glenn then went back to the funeral home.

The next day Sheriff Wilcox went to Glenn's parent's house and told them that the military was very angry at Glenn and *he was not to talk about anything he saw or he and his whole family would suffer severely.* Later that morning, the nurse Norma Marie Sels that he had seen the previous day called Glenn. She was still crying and told him to meet her at the officer's club. When they met, she looked white and was on the verge of hysterics, seemingly sick to her stomach.

She said that there had been a crash and that they had brought two small mangled bodies into the infirmary in body bags. She said that their smell was sickening and they looked horrible. She also stated that the doctor's talked about the bodies being toxic and ordered her to take notes. These doctors were flown in from Walter Reed Hospital. The nurse then described the bodies as being small with large heads and slits for mouths. They had 4 fingers with little pads on the tips of the fingers. They had no teeth or tongues, but firm cartilage instead. Their noses were concave with two openings and no bridge. Their eyes were large. Glenn Dennis never mentioned the nurse's name to the first interviewer because he feared that she would get into trouble. However, he never saw the nurse again as she was transferred to England. *Later, he heard that she had been killed in an airplane crash! Coincidence?*

## Fact: The Pilots who flew the wreckage out of Roswell

### Interviewee: Sappho Henderson

Captain Oliver Henderson, known as "*Pappy*," held a top secret clearance and was a transport pilot in the Army Air Corp. In 1981 while near death, he confided to his wife Sappho, that 34 years earlier he flew the wreckage of the crashed saucer from Roswell Army Air Field to Carswell Army Air Field. He described seeing small alien bodies with large heads and eyes in hanger 84 before they were loaded on to his plane. He was an honest man that never told a lie and kept all of the government secrets that were entrusted to him. However, knowing he didn't have long to live, he felt compelled to tell his wife the truth about the Roswell crash.

### Interviewee: Robert Sherkey

Robert Sherkey was the pilot of the B-29 bomber that flew the crash material and bodies to Wright-Patterson Air Force Base from Carswell Air Force Base. He stated that the material that General Ramey showed the press was not from the Roswell crash site, but just normal weather balloons.

## Rumor: Presidential Visitation

President Eisenhower had learned of the capture of the flying saucer and found out that some of it was at Edwards Army Air Field. He was advised that it was being studied at that location before it would be shipped to Wright–Patterson Army Air Field. He then had his staff set up a golfing vacation. He supposedly went in February of 1954. On February 20, 1954 President Eisenhower went somewhere without his entourage and rumors began spreading that the president was missing or possibly dead. Later a cover story was released that the president had lost a cap on his tooth and was getting it repaired. It was believed at this time he was viewing the saucer and its dead preserved occupants.

## *Rumor: Other Presidential Visitations*

It has been rumored that only six presidents were authorized into Area 51 to view the alien technology and its occupants, these presidents were: Harry S. Truman; Dwight D. Eisenhower; John F. Kennedy; Richard Nixon, Ronald Ragan and George Bush Sr. all others were denied access to this base. Some conspiracy theorists indicate that JFK was denied access upon first request. He was worried that the Russians and the Cubans would mistake Unidentified Flying objects as American aircraft, which could start a nuclear war. They say that he pushed the issue and was finally granted permission to view the evidence. There was also rumor that he took a certain individual with him, which eventually led to his assignation. Their evidence for this rumor is said to be in a recovered document called the "Burned Memo." **(Below are parts of the Burned Memo.)**

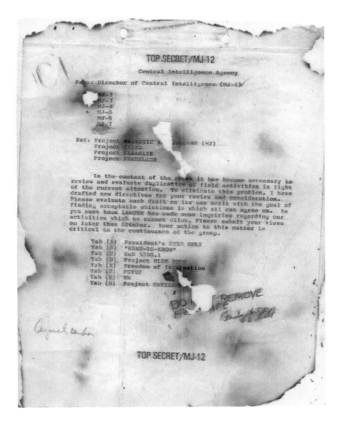

**TOP SECRET/MJ-12**

Central Intelligence Agency

From: Director of Central Intelligence (MJ-1)

- MJ-2
- MJ-3
- MJ-4
- MJ-5
- MJ-6
- MJ-7

Ref: Project MAJESTIC 12 JEHOVAH (MJ)
Project ___
Project PARASIYN
Project PARHELION

In the context of the above it has become necessary to review and evaluate duplication of field activities in light of the current situation. To eliminate this problem, I have drafted new directives for your review and consideration. Please evaluate each draft on its own merit with the goal of finding acceptable solutions in which all can agree on. As you must know, LANCER has made some inquiries regarding our activities which we cannot allow. Please submit your views no later than October. Your action to this matter is critical to the continuance of the group.

- Tab (A)  President's EYES ONLY
- Tab (B)  "NEED-TO-KNOW"
- Tab (C)  DoD 5200.1
- Tab (D)  Project BLUE BOOK
- Tab (E)  Freedom of Information
- Tab (F)  PGTOP
- Tab (G)  BW
- Tab (H)  Project ENVIRO___

TOP SECRET/MJ-12

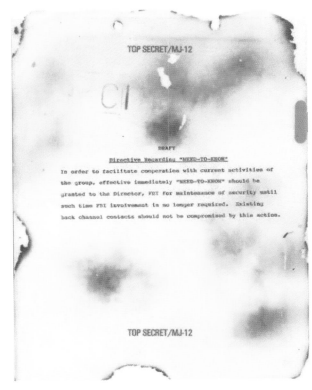

**TOP SECRET/MJ-12**

DRAFT

**Directive Regarding "NEED-TO-KNOW"**

In order to facilitate cooperation with current activities of the group, effective immediately "NEED-TO-KNOW" should be granted to the Director, FBI for maintenance of security until such time FBI involvement is no longer required. Existing back channel contacts should not be compromised by this action.

TOP SECRET/MJ-12

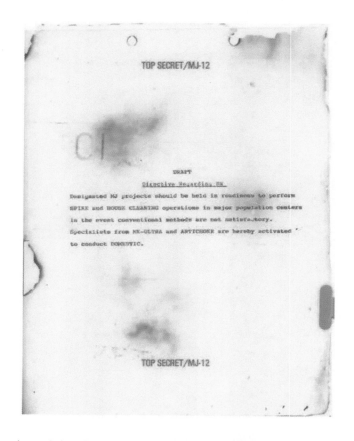

**Speculation:** A series of documents was sent to a UFO researcher Jamie Shandera on December 11, 1984 showing a hierarchy echelon that over sees any UFO data that is considered detrimental to the defense of the United States. This committee is known as MJ-12 or Majic 12 **(Majority Agency Joint Intelligence Committee),** which is a symbolism for the number of individuals on the committee. The Committee supposedly consisted of the following individuals:

Central Intelligence Director Admiral Roscoe Hillenkoetter
Secretary of Defense James Forrestal
Lieutenant General Nathan Twining of the Army Air Force
Professor Donald Menzel – Harvard astronomer and Naval Intelligence Cryptography
Vannevar Bush – Joint Research and Development Board Chairman
Detlev Bronk – Chairman of the National Research Council
General Robert Montague – Commander of Fort Bliss and White Sands Testing Grounds
Gordon Gray – Secretary of the Army and Chairman of the CIA's Psychological Board
Sidney Souers – Director of the National Security Council
General Hoyt Vandenberg – Central Intelligence Group Director

Jerome Hunsaker – Director of the National Advisory Committee on Aeronautics

Lloyd Berkner – Member of the Joint Research and Development Board.

These documents are controversial in that they supposedly have flaws in dating, numbering, ranking of individuals involved and typesetting of letters (typewriters used at that time period). However, Doctor J. Allen Hynek, who was a member of Blue Book research, admitted in his last interview that any UFO reports that effect national security went to some other location and was never seen by the researchers of Blue Book. He also admitted that they, the researchers, were told to down play all reports by the Robertson Panel in that they were to explain these sightings as weather balloons, swamp gas, birds, atmosphere anomalies and mistaken identity of known aircraft.

**Question:** *If Majic 12 is false and there wasn't a higher order of individuals controlling this research, then who ordered this committee to detour the research investigation?*

Doctor J. Allen Hynek, frustrated because he realized that the research was tainted, quit the committee and went public about Blue Book before his death. It appears that Majic 12 data may or may not be real, but perhaps the individual or individuals that sent the material knew about a council, or committee that really did control the UFO research in the government (called a satellite government) and that this was their way of exposing it. Whether they call themselves the Majic 12 or the Eleventh Hour or etc. it may well be that these are the people who decide which presidents can be trusted with the UFO secrets and set forth the policy on how the handling of UFO and Alien visitations should be performed.

## *Fact: Secretary of Secrets*

### Interviewee: Charles Welhielm – Handyman – 2nd Hand Information

Mrs. Norma Gardner, who once worked at Wright – Paterson Air Force Base in Dayton, Ohio, was now confined to her bed due to cancer. She called her handyman Charles Welhielm over to her and told him an incredible story. She said that in 1955, while working for the government and holding a top secret clearance, part of her job description was to catalogue all UFO related material. She said that over the course of her duties there where at least a 1,000 different items that the government had recovered from UFOs. During that time, she had a chance to visit a restricted area and saw two saucers like craft. One was intact, but the other was damaged. She also saw two humanoid bodies being moved from one room to another. She was then given their autopsy report. The bodies were encased in some sort of solution and were about 4 to 5 feet in height. They had large heads with slanted eyes. The handyman asked her why she was telling him this. She said that the government couldn't do anything to her once she was dead *implying that she too may have been threatened about speaking out on what she had witnessed.*

## Fact: Roswell Crash 2nd Site Continued

Sergeant Steven Arnold military police officer told Lieutenant Colonel Corso that he had been in the tower and viewed the radar screen on the night of July 3, 1947 at Roswell Army Airfield. He stated that he never saw blips behave like what was on the screens that night. The blips appeared to be moving at 1,000 mph. Then around 04:00 a.m.,` the blips vanished, however; a crash was reported just after that. At 04:30 a.m., the Sheriff of Chaves County and the Roswell Fire Department were dispatched.

Sergeant Arnold and his team arrived at the crash site first and found a group of civilians already at that location. There in plain view was a dark colored egg shaped craft with slanted wings and two tail fins. There were three dark gray figures that appeared to be dead sprawled on the ground. They were about 4 to 4 ½ feet in height. Two more beings were alive on the side of the craft, one lying on the ground while the other was walking around. Both were crying without making a sound, but he heard them in his head. It was about this time that another military group arrived, which he thought was headed by Major Marcel. Sergeant Arnold was instructed to then remove the civilians from the crash site area.

Sergeant Arnold described the beings as having large heads and small bodies. **They had human like features with large dark eyes set wide apart. They had a small nose and a slit for a mouth with indentations for ears.** Their hairless skin was a grayish brown. One of the creatures, however, tried to escape, but a military police shot him out of fear.

It was after the area had been secured that a Counter Intelligence Corp officer directed him and his men to pick up the wreckage. Then the police and fire departments arrived from Roswell and saw the military loading the bodies into the vehicles. One of the creatures was strapped to the stretcher showing signs of life, but he appeared to be dying.

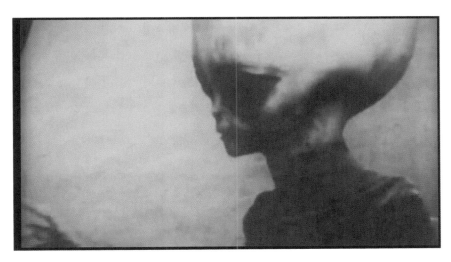

**Question:** *Doesn't this frame from a military film EXACTLY match the aforementioned description by Sergeant Arnold?*

**Analogue:** Everyone agrees that something crashed at Roswell, New Mexico on July 3-4, 1947; but there are differences about what crashed. The government first claim that it was a UFO then changed its story to balloon(s) carrying cosmic ray experiments, nuclear radiation and sound monitoring equipment and eventually added dummies to the equation. It is doubtful if a radiation monitoring balloon could pick up any radiation readings from Russia, which is over several thousand miles away. Any nuclear testing done by the soviets would have had the radiation dispersed into the atmosphere long before prevailing winds could have swept it into the atmosphere of the United States. These balloons would have only picked up normal readings.

**Question:** *If radiation was that intense, then why weren't the Philippines or Guam affected when the United States dropped the atomic bomb on Hiroshima and Nagasaki?*

The higher you go in the atmosphere the higher the concentration of radiation that filters in from space, which would make it difficult to determine if the radiation was from another government's testing or from space. An example of this can be taken from the governments own files on Lieutenant George Gorman encounter with a supposed UFO, which is contained in a later section of this book.

The Army Air Force investigators found significantly higher radiation readings on Lieutenant Gorman's P-51 aircraft after returning from his chase. These radiation readings were determined to have originated from space radiation filtering in through our atmosphere. Lieutenant Gorman was only flying at 16,000 feet. The radiation detecting balloons such as the one's used in Project Mogul reach a height from 60,000 to 75,000 feet. If at 16,000 feet the radiation was substantial, then at 60,000 feet or higher it would be significantly more and thus giving inaccurate information on any nuclear testing being conducted by a foreign government.

It was also theorized that the Soviets nuclear weapons were based on creating a greater blast ratio and less concentrated radiation fallout, which would allow their troops to invade their target at an earlier time period because the radiation would be low or almost non-existent. This was believed to be a neutron bomb that contains no uranium jacket or fission trigger, thus emanates a low radiation field. The United States Intelligence services would have known this because of this information would have been passed on through their operatives. Therefore this balloon explanation appears to be a bust (no pun intended).

**Question:** *Were all these individuals involved who had claimed to see the UFO and alien bodies some part of an elaborate government hoax or misinformation ploy or perhaps a drug-induced mass hallucination?*

**Answer:** We believe there is nothing phony about the UFO crash at Roswell because of all the separate facts that add up and support each other. The balloon story appears to be the fabrication, despite what the skeptics and the government would like the country to believe. It doesn't matter that the balloon story holds as much water as a bucket without a bottom. For the government and their skeptic allies, it's like throwing manure at a wall, hoping that some of it will stick. But when you dissected as we did, you can easily see the lies they have spread. It is obvious that their rationale is based on neither fact nor logic. And it is the right of "We, The People" to know what truly happened.

**Question:** *The Sheriff's department; the entire Roswell Fire department; a government Counter Intelligence Agent; a government Conservation Agent; a Military Pilot; a Military Nurse; a Military Intelligence Officer; a Military Secretary; a town Mortician; and several ranchers, could they all have been wrong or mistaken in what they saw?*

As you see, these are all types of trained observers and reliable witnesses, so why should all of their testimony be so blatantly discounted. It appears that the only answer that makes any sense is that these were not dummies and a weather balloon, but rather real alien bodies and the flying saucer they crashed in. If this is the truth then you must ask yourself is the United States government, the citadel of freedom, covering up the greatest event of the 20th Century?

## *Fact: Where some of the crash material went*

Lieutenant Colonel Philip Corso was a major when he was stationed at Fort Bliss, Texas. The base was located just outside of El Paso, Texas. Master Sergeant William Brown was a friend of Lieutenant Colonel Corso and was on the same bowling team. They had heard rumors about the Roswell crash and on the night of July 6, 1947; Master Sergeant Brown saw some odd crates arrive that evening. He investigated the crates and found that they were from the 509th Bomb Wing. Master Sergeant Brown immediately informed Major Corso, who was duty officer that night. They went to inspect the crates. Opening one of the crates, they found it to contain a small coffin that had a glass cover. There was a light blue gelatin type liquid that filled the coffin and they could see a 4-foot human shaped figure with 6 fingers on the hands and thin arms and legs contained within the coffin. The head was oversized with almond shaped eyes and gray skin. It had no hair and no definition to the cheeks. Major Corso said that he was quite disturbed at the sight of this being.

**Speculation & Summary:** By combing the stories of the Roswell UFO crash, a better conception of what occurred that morning of July 4, 1947 comes to light. It appears that on the previous night of July 3, 1947; the radar screens at Roswell Army Airfield came alive with unknown blips that were being tracked at great speeds. During this time, a lightning storm took place and one of these crafts, most likely containing a electromagnetic propulsion system, was struck by a bolt of lightning that blew a piece of the craft off and spread debris across the Foster ranch where the Brazel Family resided. An alternate theory claims that there were two saucers, which collided during the storm and the United States government recovered both saucers. I refer to a book written by Nuclear physicist Stanton Freidman and Donald Berliner called "*Crash at Corona.*"

There is also a third and slightly alarming possibility of two spacecraft having a "dogfight." This is deducted from no attempted being made to rescue the downed spacecraft. Whichever occurrence is correct, a motorist around 4:00 a.m. spotted a fireball coming down and crashing, he thought it was an airplane and called it in to the sheriff's department.

The sheriff dispatched his own men and Roswell fire department to stop the spread of any flames from the crash. He also notified Roswell Army Airfield. In the meantime, civil engineer Barney Barnett saw the smoke from the crash and thinking it was an airplane went over to assist. He arrived at the crash site just before a group of archeologist students who also saw the smoke and came over to investigate. Shortly later, the military arrived and took control of the crash site. They swore everyone to secrecy and sent them away just before the Chaves County sheriff and Roswell fire department arrived. They witnessed the bodies being loaded onto the trucks and ambulances by the military. The vehicles were then driven to Roswell Army Air Field where they were unloaded in hanger 84. The living creatures may have then been taken into the infirmary.

Around this time, Mack Brazel checking the ranch because of the lightening storm the previous night came, upon the 1st wreckage site. He picked up some of the material and later that day, puzzled by the properties that this material displayed, went over to his nearest neighbors Floyd and Loretta Proctor showing them what he had found. By this time the bodies were already at Roswell Army Air Field and arrangements being made to fly the bodies and crashed craft to Carswell Army Air Field in Fortworth, Texas with Captain Henderson as the pilot.

Upon reaching Carswell Army Air Field the military then decided for whatever reason to split the material from the crash and send it to several locations retaining some in hanger 18. A convoy was set up and a portion of the crash was driven to Fort Bliss in El Paso,

Texas. Another transport was set up and some material was flown to Edwards Army Air Field in California. The remaining was sent to Langley Air Field in Virginia. Later the majority of this material and bodies would be sent to Wright-Patterson Army Air Field to reside in building 18A.

On Sunday July 6, 1947, Mack Brazel informed the Sheriff of Chaves County about the debris that he had found on his property. The Sheriff now knowing about the alien crashed vehicle and bodies advised Mister Brazel that he needed to inform Roswell Army Airfield about this new crash site. Mack Brazel most likely was warned by the sheriff not to leave town until the military arrived. W.E. Whitmore of KGFL radio then encountered Mack Brazel knowing about the rumors of the crashed saucer and brought him to Walt's house for an exclusive interview.

Major Marcel now knowing of the previous alien crash and seeing the alien bodies was assigned by Colonel Blanchard to retrieve Mister Brazel. Major Marcel then went to town and found Mack Brazel. After speaking with Mister Brazel, Major Marcel had Mack Brazel accompany him to Roswell Army Airfield to meet with Colonel Blanchard. Mack Brazel then led a convoy to the spot of the first crash site. He was then escorted back to Roswell Army Airfield where he was held for several days. Upon gathering up the rest of the crash site evidence Colonel Blanchard then authorized a news release.

The material recovered at the Foster ranch was also flown to Carswell Army Air Field and stored in hanger 18. On July 9, General Ramey set up a news interview with the cover story of a downed weather balloon. It was feared that the public would panic as they did when Orson Wells broadcasted the "*War of the Worlds*" at the Mercury Theater some years earlier. President Truman realizing that these beings and craft from another world may be a threat to our civilization, assigned the director of the Central Intelligence Agency, Admiral Roscoe Hillenkoetten, to bring a committee together that would design policy and procedures on how to deal with these extraterrestrial visitations.

**The Ramey Memo**

**NOTE:** *Recently in the photograph of General Ramey at the news conference about the Roswell crash, he was holding a piece of paper, and that paper was analyzed by advanced computer imaging several times. In one of the attempts, supposedly the words "bodies, disc and crash" were brought out. However, the Ramey's document is still the subject of debate and considered inconclusive. Serious problems to consider when examining any photographs are associated with the image being an accurate reproduction of what was truly printed: in this case, on the paper in Ramey's hand. There are many factors affecting the image recording process:*

- *Resolution of the lens.*
- *Focus of the camera.*
- *Resolution of the film.*
- *The type, temperature, and duration of developing solution used.*
- *Resolution of the paper the image was printed upon.*
- *Resolution of the lens used to enlarge the image for printing.*
- *Resolution of the scanner used to digitize the image.*

All these factors add to the amount of noise present in the image. The more variables involved, the more noise is generated.

## Fact: The Material Found at the Crash Site

General Twining said in a later news interview that the strange sightings are real and not imaginary, and are approximately the size of a man made aircraft. General Nathan Twining wrote:

"2. It is the opinion that:

"(a) The phenomenon reported is something real and not visionary or fictitious.

"(b) There are objects probably approximately the shape of a disc, of such appreciable size as to appear to be as large as a man-made aircraft.

"(c) There is the possibility that some of the incidents may be caused by natural phenomena, such as meteors.

"(d) The reported operating characteristics such as extreme rates of climb, maneuverability (particularly in roll), and action which must be considered evasive when sighted or contacted by friendly aircraft and radar, lend belief to the possibility that some of the objects are controlled, either manually, automatically or remotely."

**(Below are copies of the first two pages of the Twining Memo.)**

**SECRET**

TSDIN/HMM/ig/6-4100
23 September 1947

TSDIN

SUBJECT: AMC Opinion Concerning "Flying Discs"

TO: Commanding General
Army Air Forces
Washington 25, D. C.
ATTENTION: Brig. General George Schulgen
AC/AS-2

1. As requested by AC/AS-2 there is presented below the considered opinion of this Command concerning the so-called "Flying Discs". This opinion is based on interrogation report data furnished by AC/AS-2 and preliminary studies by personnel of T-2 and Aircraft Laboratory, Engineering Division T-3. This opinion was arrived at in a conference between personnel from the Air Institute of Technology, Intelligence T-2, Office, Chief of Engineering Division, and the Aircraft, Power Plant and Propeller Laboratories of Engineering Division T-3.

2. It is the opinion that:

    a. The phenomenon reported is something real and not visionary or fictitious.

    b. There are objects probably approximating the shape of a disc, of such appreciable size as to appear to be as large as man-made aircraft.

    c. There is a possibility that some of the incidents may be caused by natural phenomena, such as meteors.

    d. The reported operating characteristics such as extreme rates of climb, maneuverability (particularly in roll), and action which must be considered <u>evasive</u> when sighted or contacted by friendly aircraft and radar, lend belief to the possibility that some of the objects are controlled either manually, automatically or remotely.

    e. The apparent common description of the objects is as follows:

        (1) Metallic or light reflecting surface.

**SECRET**

U-39552

Basic Ltr fr CG, AMC, WF to CG, AAF, Wash. D.C. subj "AMC Opinion Concerning "Flying Discs".

 (2) Absence of trail, except in a few instances when the object apparently was operating under high performance conditions.

 (3) Circular or elliptical in shape, flat on bottom and domed on top.

 (4) Several reports of well kept formation flights varying from three to nine objects.

 (5) Normally no associated sound, except in three instances a substantial rumbling roar was noted.

 (6) Level flight speeds normally above 300 knots are estimated.

f. It is possible within the present U. S. knowledge -- provided extensive detailed development is undertaken -- to construct a piloted aircraft which has the general description of the object in subparagraph (e) above which would be capable of an approximate range of 7000 miles at subsonic speeds.

g. Any developments in this country along the lines indicated would be extremely expensive, time consuming and at the considerable expense of current projects and therefore, if directed, should be set up independently of existing projects.

h. Due consideration must be given the following:-

 (1) The possibility that these objects are of domestic origin - the product of some high security project not known to AC/AS-2 or this Command.

 (2) The lack of physical evidence in the shape of crash recovered exhibits which would undeniably prove the existance of these objects.

 (3) The possibility that some foreign nation has a form of propulsion possibly nuclear, which is outside of our domestic knowledge.

3. It is recommended that:-

 a. Headquarters, Army Air Forces issue a directive assigning a priority, security classification and Code Name for a detailed study of this matter to include the preparation of complete sets of all available and pertinent data which will then be made available to the Army, Navy, Atomic Energy Commission, JRDB, the Air Force Scientific Advisory Group, NACA, and the RAND and NEPA projects for comments and recommendations, with a preliminary report to be forwarded within 15 days of receipt of the data and a detailed report thereafter every 30 days as the investi-

COPY   SECRET   U-39552

According to Lieutenant Colonel Philip Corso, President Truman asked Central Intelligence Director Admiral Roscoe Hillenkoetter and Secretary of Defense James Forrestal if the United States government should tell the American public about the Roswell Crash and the alien bodies. They replied "no" that our government had to study the material first. The autopsy suggested that the aliens were humanoid and may somehow be related to the human race on our planet. It was noted that the aliens communicated through some form of telepathy or thought projection. However, these creatures' organs, bones and skin composition were different from ours. The heart and lungs were larger than ours. Their bones were thinner yet stronger and more flexible. Their skin showed difference in atomic alignment. Their heart did not work as hard as ours thus it appears that our gravity is less than the gravity on their world. Their lungs appear to store atmosphere like a camel's hump stores water. It was noted from the two beings that were still alive had difficulty breathing our atmosphere.

**Speculation:** By the descriptions of the creatures given above one can presume that the planet these beings are from is larger than our own. Therefore, their gravity would be stronger than the Earth's and their atmosphere of higher pressure. This would account for their smaller frame and the fact that their hearts would not have to work as hard as our human hearts on our planet.

Their world must also be a little closer to their sun, which would account for their eyes having a dark filter over them and their skin being denser in atomic structure. Their planet atmosphere may consist of more Carbon Dioxide then ours, which may account for their breathing difficulty. Our atmosphere contains 78 percent nitrogen, 21 percent oxygen and 1% Carbon Dioxide with other gases. If their world had more nitrogen they would succumb to nitrogen narcosis, which would cause them to hallucinate (seeing things that aren't there – problem divers have if the mixture in their tanks is incorrect when underwater). Thus, their atmosphere must contain more Carbon Dioxide then ours as more Oxygen would not have an effect on humans.

**Note:** More Carbon Dioxide may indicate more plant life which backs up Condor and Falcon (the two intelligent agents from the TV Documentary "UFO Live") in which they state that the aliens eat only vegetables and fruits. There is also raise another possibility and question. The planet Mars has an atmosphere of 95% Carbon Dioxide. *Is it possible these beings can actually breathe the atmosphere of Mars...and may well have bases located there?*

## A blend of Fact and Unsubstantiated Account: The Material Found at the Crash Site - Continued

Lieutenant Colonel Philip Corso in 1961 was assigned to the Pentagon. He was given a position under General Trudeau in the Army Research and Development Division.

Lieutenant Colonel Corso and General Trudeau became friends while Lieutenant Colonel Corso was on President Eisenhower's staff. One day General Trudeau called Philip Corso up to his office and took him over to a set of file cabinets. He told him that these cabinets were to be moved to Philip's office. General Trudeau also stated that Philip Corso would be working in the Foreign Technology Division as part of his duties. He was also to keep track on what technology other governments were working on. General Trudeau said that the material that was in these cabinets would fall under Philip's responsibility, but that it was different from the normal material. General Trudeau purportedly then mentioned the Roswell incident and that Philip should review the paperwork within the cabinets.

In the cabinet, Philip found a shoebox with odd material inside. There was a tangled set of wires; strange cloth; a visor like headpiece and little wafer like crackers of dark gray color. The general wanted Corso to set up contracts with companies that worked for the government to find out what these things did and what we could use them for. Some companies that already worked for the government at that time were Northrop, Boeing, Ford, Bell, Sikorski, Hughes, Lockheed and McDonald/Douglas.

Corso looked at the material and tried to rationalize what they could be used for. The wires looked like it may have transmitted or conducted light. The wafers looked like a kind of small circuit board. The eyepiece seemed to reflect or amplify light. The strangest device of them all was the headband, but Corso couldn't determine what it was used for until speaking with some scientist. They theorized that since there appeared to be no control panels on the craft and that there were boards with handprints indented in them these beings must be able to communicate with their ship directly. These beings may have used the headband to transfer their thoughts through their suits to these boards. In other words, the electromagnetic impulses of their brains controlled the alien craft. They further believed that their ship ran on an electromagnetic propulsion system and that this system provided an anti gravity field and a inertia canceling system, which would protect the occupants from sudden stops and radical turns that this type of craft was known to perform. This made sense to him that the thought process of the brain directly into the craft would allow the craft to respond quicker than by manual control.

All these creatures had to do was place their hands in the indentations and the craft automatically responds to their thoughts. This was why they could out maneuver our fighters.

Lieutenant Colonel Corso said that the United States government was afraid of the EBE's (Extraterrestrial Biological Entities) were hostile and observing our military for an eventual attack. This is why the government didn't release the information about the alien visitations to the public. They feared that this would create a panic, which the soviets or the Chinese could use to their advantage. The government also tried to down play these UFO observations because they had no way of protecting the public against alien abductions. The government's concern was not unfounded for there were other examples of the aliens observing our defenses. Astronaut Gordon Cooper reported seeing a UFO landing outside of Edwards Air Force Base in 1957. This was filmed by the military and this film was sent to Washington D.C.

**NOTE:** *Gordon Cooper's actual interview can be seen in the movie "Out of the Blue."*

Lieutenant Colonel Corso cited other examples of alien observations. X-15 pilot Joseph Walker revealed that on his missions he was told to be alert for UFO activity. He stated that on one of his missions in 1962, he did observe a UFO. Astronaut Neil Armstrong reported that he had seen an alien base on the moon during the Apollo 11 mission and this observation was quelled by NASA (National Aeronautics and Space Administration). These matters were treated as high national security concerns and were brought before the National Security Council. General Douglas Mac Arthur told the New York Times in 1955 "That the nations of the world will have to unite, for the next war will be an interplanetary war. The nations of the earth must someday make a common front against an attack from people from other planets." Those in the know felt that we were facing less of an invasion and more of an infiltration, which was a matter of great concern as these aliens seemed to be testing our defenses and comprising our systems.

**Speculation:** Some critics state that Lieutenant Colonel Corso book is pure fabrication because he knew nothing about reverse engineering or research and development. In reviewing his book, he did not portray himself as having knowledge in either of these areas. Rather, he appeared as an executive assistant to the Chief Executive Officer

(CEO) of a company. The only difference is that this company is the United States Army and the CEO being General Trudeau. It is more likely that Lieutenant Colonel Corso was given an assignment to perform the initial research by speaking with scientist in different fields of endeavor. He then filed his report with General Trudeau and his suggestions that he had formed from his investigation. It is General Trudeau that then reviewed the reports and made the decisions as to the disposition of the material recovered. Whether the General conferred with those above him is not indicated. Lieutenant Colonel Corso makes no decisions on policies or procedures on his own, but merely performs the tasks that were assigned to him much like that of an executive assistant in a large corporation. Thus looking at him in this manner it is quite conceivable that he did do what he claims to have done because it takes no unique skills to perform the work assigned to him by his superiors.

## *Fact: Later Investigation*

For fifty years, the Roswell Crash is still being investigated by several creditable researchers and the following information was discovered in regards to that incident. David Soucie FAA (Federal Aviation Association) Aircraft Crash Investigator went to the crash site and measured the velocity and wind patterns/directions. There were still signs of the direction that the crash occurred, which was in a straight line according to indents in the ground. He stated because of the cross winds in that area a balloon would have blown off in an angle and not left a straight line. Whatever crash at that site was a considerable object and far more heavier than any balloon, leaving indentations 200 feet long and 10 feet in width. He also discovered that the Gamma radiation was too high for that crash site as opposed to the surrounding area, and the soil and rocks, when analyzed, were highly magnetic.

Dental Technician John Masco at the Veterans Hospital in Wright-Paterson Air Force Base said he was given a jaw bone that was larger than a human jaw bone and he tried to match it to animals jaw bones of this planet, but could not find a match. Another investigator contacted Toni Ryan a body language expert that works for the FBI (Federal Bureau of Investigation) and had her review the film of Jesse Marcel's last interview to see if he lied about the Roswell crash. She stated after watching all the film that he believes in what he said was the truth.

In going through Jesse Marcel's personal affects a diary was found and its handwriting scrutinized by Jenifer Nasso, a hand writing expert. She said that it was written by someone else. In making comparisons with the handwriting of all his work associates, she found that only one individual handwriting came the closest to that in the diary and it belonged to Roswell Air Force Base Adjacent Patrick Saunders. They spoke with Patrick Saunders' daughter and she stated that he destroyed most of the film that was taken at the Roswell crash site because the government was afraid to tell the public the truth about the alien crash at Roswell. Her father stated to her that they did not know if the aliens were friendly or not.

**NOTE:** There were 300 sightings of UFOs in 39 states that year after the Roswell crash. Astronaut Edgar Dean Mitchell stated in an interview that it really was a UFO that crashed at Roswell and not a weather balloon. Bigalow Aero Space company was said to have examined some of the material from the Roswell crash and said that it was not from this earth.

## *Unsubstantiated Account: Alien Autopsy*

By looking at the development of life here on earth, we must disregard the appearance of any aliens with abstract physiology from those of humanoids such as seen in some of the life forms portrayed in movies like *Star Trek* or books like *Beyond Mars*. In reality after reviewing known anthropology genetics, we tend to believe that intelligent alien life forms would more closely resemble our physical appearance. One should review the following section with an open mind to the possibility that alien life forms most logically would be similar to our own.

In the 1990's a producer named Ray Santilli acquired a film from an elderly gentleman, who said that he was at Carswell Army Air Field in Fortworth, Texas in 1947. This elderly gentleman stated that he was a military photographer and it was his job to make a photographic record of the alien bodies that were brought in from the Roswell crash site. He further stated that several of the aliens were alive when brought to Carswell. He allowed Santilli to run the film and after seeing the contents of this film Santilli, made a deal to purchase it from him. This film shows the autopsy of a dead female alien on a surgical table being dissected by some individuals that are presumed to be doctors. This film was then presented on the Fox network public television and contained comments both pro and con with regards to the authenticity of the subject matter.

In the film, they first reviewed the objects in the room where the autopsy was being conducted. They looked at the model telephone that was hanging on the wall and agreed that it was consistent with that of the 1947 time period. Next, they looked at the wall clock and it too was from that era in time. Finally, they reviewed the trays and utensils that the alleged surgeons were using and this too was also consistent with that time period. The producer of the show then had two pathologists review the surgical procedures being performed in the autopsy. The older surgeon Cyril Wecht (former president of the American Academy of Medical Science) agreed that the procedures being performed by the doctors on the film were standard for that time period. However, the younger pathologist disagreed with the way the autopsy was being performed.

The code on the film itself was checked by the makers of the film the Eastman - Kodiak Film Company. Padlo Cherchi Senior Curator at Eastman house and stated that the code showed that this film was produced in 1927, 1947 or 1967. Thus the film could have come from that time period. However, he also stated that he believed that the film was real because it would be too expensive to fake a film like this.

Opponents of the film said that the person doing the filming was not a real cameraman because he was not getting the best shots of the autopsy and that the camera was going in and out of focus. Radic Ryan, an actual military combat cameraman from the 1940's, viewed the film and stated that the cameraman that was doing the filming was moving around because the surgeons doing the autopsy were moving around and the cameraman was attempting to stay out of their way. Ryan also stated that the film going in and out of focus was common for the time period. This was due to the fact that the cameras of that time period did not have true to length auto-focus like the cameras of today and thus will go in and out of focus as the cameraman moves around. He also stated that he believed this film was not a fake. It should be noted that the film shown was spliced together from left over segments because of editing that took place from the actual film.

Next a Hollywood producer of horror movies reviewed the film and he thought it was a fake, however; he did not design any of the props that went into the scripts. So a special effects creator named Stan Winston was contacted, who took an entirely different view of the film. Mister Winston and his crew after looking at this film stated that this has to be real. They said that even today with the silicone material available, they cannot reproduce the peeling of the skin as seen in this film. They also stated that they could not get the blood to flow evenly as seen in the autopsy. Stan Winston stated that to try and duplicate this film of the alien autopsy today would cost hundreds of thousands of dollars, which would not be worth a studio to produce.

Some physicians reviewed the film and said that the female on the table has six fingers a sign of polydactolate and has no feminine mammary which is an indication of Turners Syndrome. Girls with this type of disorder usually die at an early age. These symptoms could be caused by a chromosome or genetic disorder.

**Question:** *Could this be the outcome from exposure to some form of radiation poisoning?*

**Speculation:** Suppose this autopsy was real and the being on the table was from another world. Some of the witnesses of the crashed space craft said that the beings only had four fingers, and yet the being on the table had six.

**Question:** *Could it be that some nuclear type catastrophe befell that planet's culture which caused some of their race to grow six fingers and toes and perhaps affected the chromosomes that gave their females Turners Syndrome?*

Perhaps, that is the reason that they are visiting this planet to obtain some sort of cure for their world. In most alien abductions it appears that these aliens are more interested in our ability to reproduce than in our intelligence or culture. Maybe their population is dwindling because of this catastrophe. During our nuclear tests, there were some animals, bugs and plants that were found in the aftermath which had abnormalities created from the radioactive fallout.

**Question:** *If it was our world that suffered from this type of situation and we had the ability to travel to other planets that had life forms similar to ourselves, would we not do the same?*

In the television show "UFO Cover-up Live," the government intelligence agents from Area 51, code named *Falcon* and *Condor* mentioned an agreement that the United States government had with the aliens from Zeta 1 Retuculi and an exchange program that had been presently in progress for at least twenty years. (Remember the statements made by Paul Hellyer at the very beginning of this book?) Is the United States government exchanging reproductive and medical information for alien technology? **(Below a Copy of photograph of the supposed "Alien Autopsy.")**

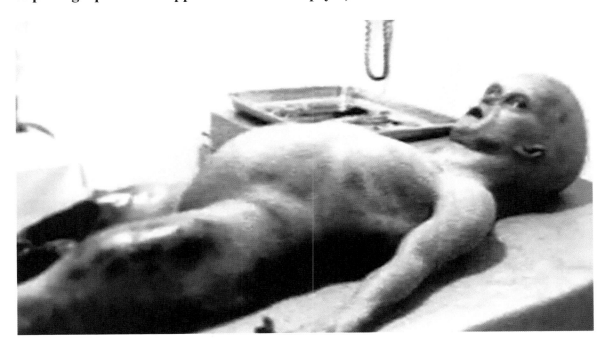

## Unsubstantiated Testimony: Chase Brandon, C.I.A. Operative

In July of 2012, a CIA veteran of 35 years by the name of Chase Brandon, spoke out on the 65th anniversary of the Roswell Incident to reveal a *supposedly* hidden CIA file on the 'UFO' that was supposedly found at the New Mexico crash site - and said, "*It really happened.*"

Chase spent his final 10 years with the agency as a official liaison on the director's staff. During this time as a director (in the mid 1990's), Brandon (*according to his account*) had walked into Langley's Historical Intelligence Collection section and began browsing through the various files. The Historical Intelligence Collection, a tightly secured and restricted area, allowed very few CIA personnel to have access to the vault.

As Brandon shuffled through stacks of boxes he noticed one box with a single hand-written word on it – "Roswell". With curiosity getting the better of him, he began examining the contents of the box.

Although he would not disclose specifically the contents of the box (his security oath prevents disclosure), what he saw changed his perception of the Roswell incident.

He since has made the following statements:

*Statement # 1*

*"It was not a damn weather balloon — it was what it was billed when people first reported it. It was a craft that clearly did not come from this planet, it crashed and I don't doubt for a second that the use of the word 'remains' and 'cadavers' was exactly what people were talking about."*

*Statement # 2*

*"Some written material and some photographs, and that's all I will ever say to anybody about the contents of that box. But it absolutely, for me, was the single validating moment that everything I had believed, and knew that so many other people believed had happened, truly was what occurred."*

**(Sources: Daily Mail, Wikipedia, Christian Post)**

He is now convinced beyond a doubt that an extraterrestrial craft crashed landed in Roswell, New Mexico in July, 1947.

Earlier publicly released documents appear to back up Brandon's story—or at least the idea that government authorities covered up involvement with aliens. One memo that appears to prove that New Mexico prior to 1950 has been published by the FBI.

The bureau has made thousands of files available in a new online resource called "The Vault." Among them is a memo to the director from Guy Hottel, the special agent in charge of the Washington field office in 1950.

In the memo, whose subject line is *'Flying Saucers,'* Agent Hottel revealed that an Air Force investigator had stated that "*three so-called flying saucers had been recovered in New Mexico.*"

The investigator gave the information to a special agent, he said. The FBI has censored both the agent and the investigator's identity. Agent Hottel then went on to write:

"*They were described as being circular in shape with raised centers, approximately 50 feet in diameter. Each one was occupied by three bodies of human shape but only 3 feet tall,*" he stated.

The bodies were 'dressed in a metallic cloth of a very fine texture. Each body was bandaged in a manner similar to the blackout suits used by speed flyers and test pilots.

**CIA headquarters at Langley, Virginia where Chase Brandon claims to have seen a secret room in which the 'truth' about the Roswell incident is kept**

**Chase Brandon worked for the CIA for 35 years, and has overseen covert operations in 70 countries.**

## A Roswell Staff Officer Claims There Was A Cover-Up

Lieutenant Walter Haut, who was the public relations officer at the base in 1947, and the man who issued the original and subsequent press releases after the crash on the orders of the base commander, Colonel William Blanchard, died in 2006, leaving the sworn affidavit to be opened only after his death, as previously mentioned.

The text asserted that the weather balloon claim was a just cover story, and that the real object had been recovered by the military and stored in a hangar. He described seeing not just the craft, but alien bodies.

Haut's affidavit talks about a high-level meeting he attended with base commander Colonel William Blanchard and the Commander of the Eighth Army Air Force, Gen Roger Ramey. Haut stated that at this meeting, pieces of wreckage were handed around for participants to touch, with nobody able to identify the material.

He said the press release was issued because locals were already aware of the crash site, but in fact there had been a second crash site, where more debris from the craft had fallen.

Haut also spoke about a clean-up operation, where for months afterwards military personnel scoured both crash sites searching for all remaining pieces of debris, removing them and erasing all signs that anything unusual had occurred.

"But it absolutely, for me, was the single validating moment that everything I had believed, and knew that so many other people believed had happened, truly was what occurred."

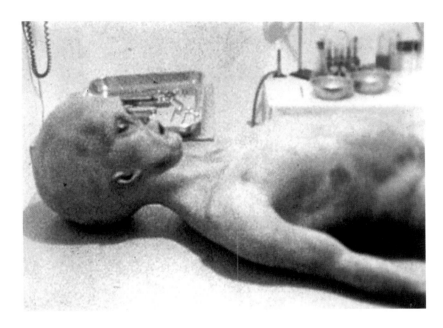

**NOTE:** This particular photo above was supposedly taken during alien autopsy and is said to be archived in the CIA files.

# Chapter V: Alien Abductions,

## Teleportation VS Quantum Physics

### *Rumor: Attempted Teleportation*

In 1954, it was rumored that the United States government experimented with a teleportation project. Their premise was that if the experiment was successful, they could teleport U.S. troops anywhere on the planet at a moment's notice. The initial locations for the test were supposed to be the starting point of Langley, Virginia (Central Intelligence Agency Headquarters) and ending at San Diego, California (unspecified location). The first item said to be sent by "*beaming*" was a wooden chair. The estimated time of arrival was to be 5 minutes. Rumor states that this part of the test was successful. The chair arrived intact. The second phase was to beam a plant because it represented a living organism. The plant did arrive, but not quite in the same condition as it initially started. The third phase was the critical one. For this phase, a poodle was chosen. The results were not what they expected. What reassembled before their eyes bore no resemblance to a dog and lived for only a few seconds. It was determined that inanimate objects could be beamed, but living objects could not. However, remember ***this is only a hearsay account.***

### *Fact: Quantum Teleportation*

It is fact that both European and Chinese Physicists have *Quantum teleported photons* using lasers over 88 and 100 miles respectively in the recent years. However, the process used is not something that disappears and reappears like in **Star Trek**.

Instead, the information contained in the photon's quantum state is transmitted from one photon to another through quantum entanglement. In other words, the photon doesn't actually travel the intervening distance…or instantaneously for that matter. That's because the transfer of information occurs when the sender measures the quantum state of their photon, which causes the receiver's entangled photon to instantly change.

However, in order to understand the information, the receiver has to know what the original measured information was, along with some other instructions. Those instructions are sent via normal communications, which are limited to being slightly slower than the speed of light.

It has long been theorized by science fiction novels, movies and television series that humans can be teleported through walls, buildings and even space to another location. The **Star Trek** series was especially known to use this method of transportation to visit other planets. They would have their crew step onto a transporter platform on their space ship, *Enterprise*, and beam their physical being to the planet's surface below. This teleportation theory was first conceived as a byproduct linked to physicist Werner Heisenberg.

Werner Heisenberg 1901 – 1976 was a renowned physicist who made major contributions in the field of atomic structure. In 1935, he developed the theory of quantum mechanics in which the mathematical formula for atomic systems is based and created the Theory of Heisenberg Uncertainty Principle. Applied to quantum teleportation the *Heisenberg Uncertainty Principle* simply states that the physical presence and physical movement of atomic structure cannot be measured by the same computer.

In layman's terms, this means that in the atomic structure of a living being, the particles that make up that being, have a designated location in that being. However, because all matter is condensed vibrating energy, the particles are always vibrating in conjunction with particles around it. Thus, the computer to transmit this life form must capture both movement and location *of every particle* that makes up the individual that is to be teleported. This cannot be done for the reason that computers can only see the movement or the location of these particles, *but not both*. Therefore, inanimate objects, *which are not alive*, may be able to be teleported since their physical structure may only contain micro flaws out of sync with adjacent particles. However, living beings cannot be teleported since such flaws would be life threatening...if not deadly.

It is theorized by some scientist that the particle movement and the particle location of the individual can be separated at the point of teleportation then reassembled at the final destination. However, this presents a new set of problems.

**The first problem** is that the breakdown of the particles on the individual and the resemblance of these particles must be performed at or close to the speed of light. A quantum computer (*a computer that can perform more than one function on the same subject at the same time*) would be needed to perform this function. This is a computer that has the ability to pick up the smallest amount of radiated matter by the varying of the frequency and the vibration. In other words, it must break up the data stream of a living individual into binary code (0s and 1s), transmit and reassemble that code simultaneously. It must also distinguish inanimate material (clothing, jewelry, weapons, etc.) from the individual being teleported and reassemble that material on the individual without making it part of that individual.

**The second problem** is in the reassembling of the individual. Unlike *Star Trek* where they can teleport an individual to any location on the planet, a second quantum computer would be needed to receive the data stream of the individual being teleported else the particles of that individual would be lost in the environment. This also applies to inanimate objects as well. The best example of this is a fax machine. If you send a fax to another location there must also be a fax machine synced with the first at that location to receive that fax...*else that information is lost!*

**The third problem** is that of the environment that surrounds the individual being teleported. The environment must be sterile as an operating room. Any outside matter contaminating the space could affect the re-assemblance of the individual being teleported. An example would be cobalt programming in that the slightest misplacement of any punctuation could stop or change the outcome of the program. The same would hold true in teleporting a data stream of an individual in that if an outside particle or insect were to enter the teleportation field in either the dissecting or reassembling stage of the individual being teleported, then this outside particle could interrupt the outcome of that teleportation. In other words, the individual maybe reassembled with his nose on his forehead or a fly's head. Therefore, this sterile environment must exist at both ends of the teleportation field.

**The fourth problem** is that of a theology sense…one of consciousness and one of the soul. The question that presents itself is in the living individual that was dissembled at point A.

**Question:** *Is that individual the same person that was assembled at point B? It is a question of the memory, personality, consciousness and spiritual being. Will they be the same? Theologians may argue that the soul of this individual may not have been teleported with the individual…and they just may be right!*

**Answer:** If you breakdown a person into particles (and remember, in true Quantum Teleportation you are only transmitting *the information* about that particle and *not the particle itself*), you would, therefore, be killing off the original…while creating only a pseudo carbon copy at the other end.

## *The fifth problem represents a series of questions.*

**Question:** *How will teleporting the individual affect that person's health? Will the individual's internal organs be damaged?*

**Question:** *Will that individual feel any pain when being disassembled?*

**Question:** *If the individual feels pain, will it be of such intensity as to kill that individual being teleported?*

**Speculation:** We believe that the original person's body would be turned into a glob of molecules or simply vaporized due to the process. Any depending on the speed it's performed at, there might be considerable pain. We believe they would absolutely die.

It is for this reason that any individual claiming to be abducted by aliens via teleportation has to be closely looked. Skeptics have seized upon this and have concluded teleportation of abductees must be disregarded as the product of hallucinations induced by rim sleep *(that sleep which is on the edge of the sleep cycle before wakening)*, delusion brought about by the use of some medication, paranoia induced subconsciously by some previous traumatic experience or the outright twisting of the truth in order to extract monetary gains or for their 15 minutes of exposure to the public to inflate their egos. And seemingly the Heisenberg Uncertainty Principle when applied to alien abductions suggests that aliens are not capable of teleporting humans through walls, buildings or spaceships.

However, there are two things which cannot so easy be discounted: the first is that all these abductees have described the exact same phenomena. This is interesting…and a possible clue. When it was first reported by people all over the world, it was NOT publicized due to the shame and trauma that accompanied such experiences. These people were terrified at the exposure of their ordeals…and went to great lengths to hide them from public disclosure for fear of ridicule…and what it might cost them.

The second is an incident, which occurred in the Bermuda Triangle. The incident, which we refer to, is the one involving Bruce Gernon, a pilot and real-estate broker. On December 4, 1970, Bruce flying a Beech Craft Bonanza A36 (a propeller aircraft capable of only 195 mph maximum speed) flew into some sort of cloud tunnel developing a green mist around his aircraft, where he travel over 100 miles distance in about 20 seconds of time. (This story may is well documented in the books "*The Fog*" and "*Without a Trace*.")

Bruce's airplane vanished from radar only next to reappear over Miami several minutes later (the flight was from the Bahamas to Florida). When he finally landed, he found he had ten extra gallons of fuel as well as discrepancy of 30 minutes less of flight time. By FAA rules, this had to be recorded into the aircraft's flight log, *which cannot be falsified unless the pilot wishes to lose his license*. And by these facts, Bruce's aircraft had to have been traveling in excess of 2,000 mph. This meant he was teleported or entered some sort of time vortex…and live to tell about. Something similar happened to an old-time TV/radio host, Arthur Godfrey, as he too flew through the Bermuda Triangle, saying "all time ceased."

And there was also one other pilot who experienced this green mist surrounding his aircraft while his compass spun wildly as he flew from Cuba. You may have heard of him. His name was Charles Lindbergh. When Charles Lindbergh was making a nonstop flight from Havana to St. Louis, his magnetic compass started rotating while his Earth-inductor-compass needle jumped back and forth erratically. This personal tale was revealed in his autobiography. So there are three conformations of possibly of not only magnetic phenomena, but one of possibly teleportation. A good analogy in a descriptive sense is opening a doorway to walk into a corridor, and then, opening another doorway to step out of it. Einstein had a term for this; he called it a "*Rosenbridge.*"

Now if we look into the context of abductees, they are describing to the best of their ability *their perception of a very traumatic experience.* Maybe this type of teleportation method was used instead of *Quantum Teleportation*. Lacking any further evidence or facts, particularly technological ones, it is simply ludicrous to dismiss such experiences as delusions of the mind. What seems to be the only delusion here are those in the skeptics' minds. The skeptics were not there…and do not know what really happen,…end of story! But you think a person that espouses the brilliance of their own intelligence would at least be open to the possibility of alien abductions via teleportation. Such a method would be quick and stealthy…the hallmark of any professional kidnapper.

However, about these types of abductions, we simply don't know. And since Einstein, we have been amending our standard model of physics as well as challenging and changing many of Einstein's precepts. That is fact! Who knows what will be found regarding teleportation in the future.

## *Fact: The Abduction of Betty and Barney Hill – Case # 1*

### Interviewee: Betty Hill and Marjorie Fish

Most Ufologists and readers of material pertaining to Unidentified Flying Objects are aware of the names of Betty and Barney Hill. Their names are synonymous when discussing alien abductions.

They made national if not international news when the story of their abduction became public. They were the first recorded alien abduction in the modern era.

It all started for this inter-racial couple on September 19, 1961. They had been on vacation in Canada and were driving back to their home in New Hampshire. It was after midnight and they were on highway Route 3 near Portsmith, New Hampshire by the White Mountains, when an extremely bright star caught Barney's attention when he stopped the car to walk the dog. The highway was deserted and this star appeared to be growing larger as they traveled, it was almost as if the star was moving with them.

Barney told Betty about the star and they both began to watch the star as it grew brighter. Barney eventually pulled the car over and grabbed a pair of binoculars. They both left the safety of the car at this point with Barney zooming in the binoculars while stepping away from his vehicle. It soon became clear to him that this was not a star at all, but some sort of strange craft. He could see windows on the craft and what appeared to be figures in those windows. He became frightened and ran back to the car yelling for Betty to get back into the car saying, "They mean to capture us!" After they got into the car they sped home.

When they arrived at their home they were puzzled because there were questions that they could not answer. Both their watches had stopped. They should have arrived home two hours earlier than they actually did, which meant that they had two hours of missing time they could not account for. Betty's dress was torn and the tops of Barney's shoes were scuffed. There were also highly polished scrapes on the hood of their car. Soon after this occurrence, Barney developed an ulcer and Betty began having nightmares.

For two years they suffered through these maladies. Betty then consulted a psychiatrist friend of theirs, who suggested hypnosis therapy. Finally, they went to see a psychiatrist named Benjamin Simon, who specialized in regressive hypnosis. He saw each of them separately for several months and made recordings of each session. After completing the therapy, Doctor Simon called them into his office and played back their sessions. They were astounded by what the recordings revealed, but it answered all their questions.

Hidden deep in their subconscious was the explanation of what happened to them that night of September 19, 1961. After seeing the craft, Betty and Barney got back into their car and began speeding away. However, Barney, apparently under telepathic suggestion from the aliens, turned the car down a side road. Their car's engine then stalled out and five aliens that had been in the road approached their car on either side. They opened the doors and took Barney and Betty from the car and escorted them to their space craft that was in a field not far away. Betty, extremely frightened, began to struggle and her dress was ripped during her struggles. (*Later tests on Betty's dress discovered enzymes of unrecognizable and of unknown origin stained upon it,...and that my friend is called physical evidence.*)

Once they had the couple inside the ship, they put them in separate examination rooms placing them on tables. The aliens checked the couple's bodies drawing blood also taking hair and skin samples. They checked their teeth and drew semen from Barney. They stuck a needle into Betty's abdomen and she screamed in pain. She cried out for an explanation and the alien in charge, whom she called the leader, told her that it was a pregnancy test. Betty told him that there was no pregnancy test on earth like that.

**NOTE:** Almost ten years later, our doctors develop two tests to check for pregnancy that exactly mirrored the aliens' examination of Betty Hill. These tests are amniocentesis and laparoscopy. With amniocentesis, a needle is inserted into the female's womb to extract amniotic fluid, while laparoscopy is used to examine the internal organs of the abdomen.

After her examination, Betty was taken to a larger room where she was shown a three dimensional star map. Pointing to his base star, the leader asked her if she knew where our sun was on the star map. She said that she did not know. The leader said that she simply did not have the knowledge to understand. Betty described these beings as looking similar to us, but smaller with larger heads and insect or almond type eyes. Their noses were small as was their mouths with only indentations on the side of their heads for ears.

Betty and Barney Hill were reunited and taken back to their car. The aliens inserted a mental suggestion (amnesia) to forget the abduction portion of their experience. After their sessions with Doctor Simon, Betty and Barney went to Peas Air Force Base and requested information on any unusual radar readings for the night of September 19, 1961. They received a copy of the radar report showing that the base had tracked a UFO at 02:14 hours on that night.

Betty drew a picture of the star map while under hypnosis. Sometime later, a group of Ufologists were alerted to this incident and exposed it to the press. An author named John G. Fuller interviewed them and wrote a book about their abduction entitled "*The Interrupted Journey*," which was published by Dial Press in 1966. This book included the star map. Upon hearing about the abduction an amateur astronomer named Marjorie Fish wanted to see if their story was true and went about constructing a three dimensional star map. She expected her star map would show several matching locations. Using the Gliese Star catalogue and after 7 years of work, in 1973 she found only one exact match. The base stars were located in our earth's southern hemisphere and known to us as Zeta 1 and 2 Reticuli. These star systems are 5 million years older than our sun and are classified G2 stars. The other stars in the star map which were visited by the Zeta 1 Reticulians were identified as follows: Alpha Mensae, Eridani 82, Gliese 59.2, Gliese 95, Kappa Forasc, Phi 2 Ceti, Tau Ceti and Zeta Tucanae all of which fall into the G spectral except Phi 2 Ceti, which is a F 8 spectral star.

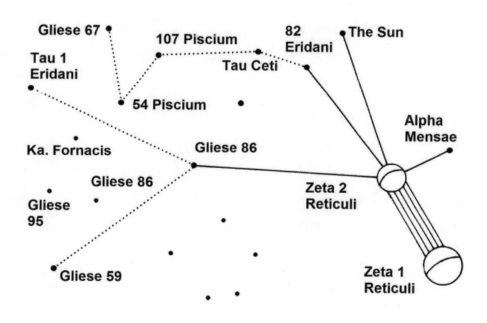

**Betty Hill's Map**

Renowned astronomer Walter Mitchell, professor of astronomy of Ohio State University, looked at Marjorie Fish's star map *and said that it was 100% accurate!* Renowned Nuclear Physicist Stanton Friedman praised Marjorie Fish for her star map creation in the movie "*UFOs Are Real*" saying that this star map verifies the Hills story and that it really happened.

**Speculation:** From the following overwhelming evidence presented it is conceivable that Betty and Barney Hill's abduction in reality did occur. First there was the radar report from Peas Air Force Base that confirms the tracking of an Unidentified Flying Object at the time of their abduction. Second is the hypnosis therapy that they both went through separately, yet their sessions confirmed what each other reported. Then there was the physical evidence: Betty's torn dress, Barney's scuffed shoes, both of their watches had stopped at the same time and the highly polished scrap marks on their car. They were both given a polygraph test (lie detector test) in which they both passed. A major piece of evidence was the star map that Betty drew under hypnosis, which clearly identifies a particular region of space in our southern hemisphere. The most significant evidence is the medical procedure of checking for pregnancy.

**Question:** *How could Betty describe a medical procedure that wouldn't be invented by human medical science until over 10 years later?*

**Answer:** Obviously, their retold experience…*was the absolute truth!*

Then the Hill's description of the aliens themselves, which matched the descriptions of the bodies found at the Roswell, also validated the said crash at Roswell. However,…the description of the bodies at the Roswell Crash wasn't released until the 1978 interview with Major Marcel when he went public with his story and researchers began checking into the crash in 1947.

**Question:** *How did Betty know where the individual stars were that these aliens traveled to unless she viewed their star map without having no knowledge of astronomy herself.?*

## *Fact: Small Aliens Reported in Newspaper Column in 1955*

On February 15, 1954, Dorothy Kilgallen, a well known and respected investigative reporter at that time, commented in her syndicated newspaper column, "*Flying saucers are regarded as of such vital importance that they will be the subject of a special hush-hush meeting of the world military heads next summer.*"

On May 22, 1955, Kilgallen also reported, "*British scientists and airmen, after examining the wreckage of one mysterious flying ship, are convinced these strange aerial objects are not optical illusions or Soviet inventions, but are flying saucers which originate on another planet. The source of my information is a British official of Cabinet rank who prefers to remain unidentified. We believe, on the basis of our inquiry thus far, that the saucers were staffed by small men—**probably under four feet tall**. It's frightening, but there is no denying the flying saucers come from another planet.*"

This article, which was separate from Kilgallen's regular column, appeared on the front pages of the *New York Journal American*, the *Cincinnati Enquirer*, the *Washington Post* and other newspapers. It even made the magazine, *Flying Saucer Review*, which alleged the information was given to Kilgallen by none other than Lord Mountbatten at a cocktail party, but attempts to verify this were unsuccessful. Although the article did not describe in detail what "*small men under four feet tall*" looked like exactly, it does verify the correct height of the aliens described in both the Roswell crash and by the Hills, *thereby validating both stories!*

**Question:** *How could the Hills describe this identical species of alien, which was also found at the Roswell crash site since that information was not known in 1961? And how did Dorothy Kilgallen know that "saucers were staffed by small men—probably under four feet tall" from presumably talking from Lord Mountbatten in 1956?*

**Answer:** Obviously, the crash at Roswell…*was also the absolute truth*. And the very next time some arrogant skeptic or brash Air Force officer infers or insists that the crash at Roswell was just a weather balloon, please enlighten them to the above stated facts.

**Question:** *Right about now, don't you think the skeptics are looking a little bit stupid if not a bit narrow-minded?*

### Fact: Travis Walton – Case # 2

**Interviewee: Travis Walton and Michael Rodgers**

In November of 1975, Crew chief Mike Rodgers and six other men, which included Travis Walton, were contracted to clear the undergrowth out of the Sitgreaves National

Forest located in Arizona. They had been working all day and dusk was quickly approaching. Mike Rodgers decided that they should quit before it became dark. They then climbed into their pickup truck and with Mike Rodgers behind the wheel they headed out of the forest. They were driving down a dirt road and entered a clearing around 16:00 hours when they saw a UFO hovering above the tree line in front of them about 100 feet away.

For some reason Mike Rodgers (whether from shock or by the Alien's mental suggestion) shut off the truck engine. It was at this point that Travis Walton got out of the truck to get a better look at the craft, which was surrounded by a golden light. He started towards the craft, thinking that it would fly away. The others in the truck were yelling for him to come back. Suddenly, a bolt of blue light struck him and prevented Travis from moving. It began to lift Travis into the air. The men in the truck were screaming at Mike Rodgers to start the truck and flee the area, thinking that Travis was dead. Mike Rodgers then started the vehicle and headed back the way in which they had come.

**Travis Walton has not been the only person to have been *supposedly* abducted in this manner. A few other "abductees" (verified though hypnosis and polygraph tests) have also verbalized strikingly similar accounts.**

It was then that they saw the golden light through the tree tops moving away from them. Having a sense of guilt overcome them, they decided to go back and look for Travis. After searching the area that the incident occurred and not finding Travis, the crew became frightened and they notified the sheriff's office. They told the sheriff's office about the incident and that Travis was missing or possibly dead. The sheriff's office then sent out a search party, which combed the area for a day and a half. Travis Walton's brother then pushed the local authorities to have the search extended for another day. However, the results were still the same, Travis was nowhere to be found.

Mike Rodgers and his crew were then brought in for questioning and not believing their UFO story, the Sheriff had each of them undergo a polygraph test because they were suspected of killing Travis Walton. Cy Gilson, from the state board of polygraph, was assigned to perform the test. He didn't believe their story *at first*, but changed his mind when five out of the six men tested had passed the examination and the sixth man's results were inconclusive. Edward Gleb President of the American Polygraph Association stated in the movie "*UFOs Are Real*" that the odds were one million to one of six people fooling a polygraph on any one subject and that the polygraph had an accuracy of 99%. Although the polygraph main focus was on whether the crew had murdered Travis Walton, Cy Gilson really believed their story.

Five days later, Travis woke up on the highway with a silver disc hovering over him. After the disc flew off, he staggered down the highway until he came upon a gas station. Unable to reach his brother by phone, he called his brother in law, who then contacted Travis's brother. They both drove down to the gas station and found Travis dazed. They told Travis what had happened and Travis thought that he had only been gone for a day, but when Travis got home he found that he had a five day growth of beard on his face *and had lost ten pounds*. Travis tried to recall what had happened to him during his missing time. Eventually, his memory began coming back to him and then realized the full scope of his encounter with the Unidentified Flying Object.

Travis after being hit by the blue beam of light recalls waking up in what he thought was a hospital room. He was having trouble breathing and realized that the air was thick.

**NOTE:** When *carbon dioxide* is present in the air at concentrations of near 600 parts per million, the air begins to feel stuffy, thick and slightly difficult to breathe. At this level, which is only 250 parts per million above your normal tolerance, serious symptoms and permanent damage are unlikely to occur.

However, as the concentration of carbon dioxide in the air approaches 1,000 parts per million, classic symptoms of $CO_2$ poisoning begin to occur. These include difficulty breathing and shortness of breath, as well as increased heart rate, headaches, hyperventilation, fatigue, sweating and possible impairment of hearing.

Here is a more technical explanation. The atomic number for oxygen is 15.9994 and the atomic number for nitrogen is 14.007, which mean that oxygen is heavier than nitrogen. The atomic number of carbon dioxide is about 44, which means that carbon dioxide is heavier than both oxygen and nitrogen. Our atmosphere is 78 % nitrogen, 21% oxygen and about 1% carbon dioxide mixed with other gases. Carbon dioxide a colorless, odorless gas is used by green plants for photosynthesis, which is the process that plants use to form carbohydrates. Thus, Travis Walton having difficulty breathing because the air was thicker means that the air he was breathing could have had a higher degree of carbon dioxide content. Thus, it is possible that the alien world contains greater amount of vegetation. (*This is also in context about was stated earlier in this book about the aliens of Roswell breathing an atmosphere with a higher carbon dioxide count on their world.*)

Suddenly, Travis became aware of three small creatures standing around him as he sat up on the table on which he laid. These creatures had large heads and eyes with small noses, mouths and ears. Climbing off the table, he began yelling at them to back away. He then grabbed a cylinder off the wall and threatened them with it. These creatures then left the room and Travis waited a little while before he too exited the room.

Emerging from the room he was in, Travis then wandered down a corridor and found a room on the right that was opened. There was a chair in that room with a large window or screen that viewed space. After entering this room, he sat in the chair and touched a lever which caused the view of space to move. It became apparent to him that he was viewing a three dimensional star map. Then he heard some noise behind him. Turning he saw a normal athletic looking human enter the room. He was wearing coveralls like the creatures, but had a helmet on. Travis went up to the man and tried to ask him some questions, but the man did not answer him and only smiled. The man then took Travis by the arm and escorted Travis through a hanger deck. In the hanger deck, Travis saw other crafts like the one he had seen in the forest.

After going through the hanger deck, they entered a small room where Travis saw two normal men and a normal looking woman waiting. They too were also dressed in uniforms, but without helmets. They tried to get him onto another examination table, but Travis began to struggle with them. Finally, one of them was able to get a mask over Travis mouth and nose. He then lost consciousness at that point. It was then he remembered waking up on the highway with the craft hovering over him.

Eventually, Travis Walton was examined by two doctors. They found nothing wrong with him other than a red spot on his right elbow, which suggested a needle had punctured the skin. Curiously, there was no acetate in his urine which should have been there if his body had gone without nourishment for 24 hours. Perhaps the needle puncture could account for this. Travis Walton went under hypnosis regression therapy and relayed virtually the same story about his abduction. Travis Walton was then given a polygraph test, but failed because he was in a highly agitated state as three psychiatrists had stated, which made that test meaningless. Several months later, Doctor Leo Sprinkler of the University of Wyoming reinstated a second hypnosis regression therapy session and polygraph test to Travis Walton, which he passed. Cy Gilson also gave Travis Walton a third polygraph test of which he also passed.

**Speculation:** Travis Walton's story has an overall essence of reality, but with some questionable elements. Why would Mike Rodgers turn off the engine of the pickup truck? In most UFO cases the electromagnetic energy of the craft usually cuts the engine off on any vehicle.

**Question:** *Did the aliens induce by mental suggestion to Mike Rodgers to turn off the vehicle?*

The beam that hit Travis was described as being blue in color. This color has been related to that being seen with an electromagnetic field. On the History Channel's television series "UFO Hunters" and on "Alien Engineering" they demonstrated that it was possible to lift a frog up with a blue colored diamagnetic field (by changing the polarity of the frog). That is to say, if the alien ship was a positive magnetic field and Travis had a positive magnetic field by changing the polarity of his magnetic field to negative, they could in theory pull him into their craft since opposite poles attract and like poles repel. Thus, the smaller object would be drawn to the larger object or repelled by the larger object. The shock of having his polarity changed may have caused unconsciousness to occur. This may be mistaken for teleportation if the individual is in a trance. or in a unconsciousness state, which may account for other abductees saying that they were teleported on to an alien craft.

**Question:** *What about the lack of acetate in Travis's urine **after he had lost ten pounds?** Was the needle mark in his arm a method for the aliens to supply some type of nourishment to his body while aboard their vessel?*

It was rumored that an American astronaut Neil Armstrong reported to NASA (National Aeronautic and Space Administration) that he had visually observed an alien base on the moon and that NASA had buried this report from the American public. If this story is true, then could this base have been where these aliens had taken Travis Walton.

The atmosphere in this alien craft Travis said was heavier and harder to breathe, which could point towards a more carbon dioxide enriched environment. The aliens in the Roswell crash also had trouble breathing our atmosphere. Their planet atmosphere, therefore may consist of higher amounts of carbon dioxide then ours, which could account for their (the aliens) breathing difficulty on Earth. This would be equivalent of Hypoxia (altitude sickness) or AMS (Acute Mountain Sickness) on our planet. (Hypoxia is a risk that mountain climbers must be aware of when reaching extremely high altitudes where the atmosphere becomes thin and causes difficulty breathing and can lead to death.) However, this information was not released until 1978 and Travis Walton's abduction occurred in 1975. Thus he would not have known about the alien's breathing difficulty.

**Question:** *How could Travis Walton have known about the heavier atmosphere difference unless he was aboard an alien vessel?*

Travis Walton did fail the first polygraph examination, which three psychiatrists claim was due to stress. However, he did pass two latter polygraph tests. What are unknown were the questions asked during these tests and not only those questions presented to Travis Walton, but those questions that were asked to Mike Rodgers and the other five men. It would be interesting to have an expert in body language known as Micro Expression Analysis (which is used by the Mossad (*Israeli interrogation method used by their police*), Secret Service, FBI and Homeland Security) were to review the film of Travis Walton's interview to see if Travis Walton or Mike Rodgers were attempting to distort the truth.

**Question:** *It is a wonder why no one has thought to have had this procedure performed?*

## Fact: The Pascagoula Abduction – Case # 3

The skeptics thought they would have a field day with this one when it first came out,...but it didn't turn out the way they envisioned. Forty years later, there is more than overwhelming evidence that it actually occurred. The Pascagoula Abduction took place in 1973, when co-workers Charles Hickson and Calvin Parker (both of Gautier, Mississippi) claimed to have been abducted by aliens while fishing off of a pier in Pascagoula, Mississippi. Like the Hills case, this incident received national media attention, and is one of best unknown cases.

It started on the night of October 11, 1973; when two local shipyard workers took in a routine evening of fishing off a pier on the west bank of the Pascagoula River in Mississippi. It had turned dark, and the two men soon heard a whizzing/buzzing sound behind them.

Both turned and were terrified to see a ten-foot-wide, eight-foot-high, glowing egg-shaped object with two blue flashing strobe lights hovering just above the ground about forty feet from the river bank. Both men, frozen with fright, then observed a door opening on the object, and three strange beings floated just above the river towards them.

They were about five feet tall, had bullet-shaped heads without necks, slits for mouths, and where their noses or ears would be, they had thin, conical objects sticking out, like carrots from a snowman's head. Although they had legs, they did not walk or move their legs at all. They had no eyes, grey, wrinkled skin, round feet, and claw like hands.

Two of the beings seized Hickson. However, when the third grabbed Parker, the teenager immediately fainted from the trauma. Hickson claimed that when the beings placed their hands under his arms, his body became numb. He was then floated into the UFOs interior and a brightly lit room. Hickson alleged he was subjected to a medical examination by an eyelike device, which also was levitating in mid-air.

At the end of the examination, the Beings simply left Hickson floating, paralyzed. They then went to examine Parker, who, Hickson believed was in another room. About twenty minutes after Hickson had first observed the UFO, he was floated back outside the craft and released. There, he found Parker weeping and praying on part of the pier near him. Moments later, the object rose straight up and shot out of sight.

Fearing ridicule, Hickson and Parker initially decided to keep quiet about their abduction, but then decided it was too important. They immediately telephoned Kessler Air Force Base in Biloxi. A sergeant there told them to contact the sheriff. But uncertain about how their bizarre story would be taken by the police, they instead drove to the local newspaper office to speak to a reporter. When they found the office closed, Hickson and Parker finally felt they had no alternative, but to talk to the local sheriff.

Understandably terrified, the men reported their astonishing story to the local police department – which understandably believed the men were simply drunk. Hickson admitted he had taken three shots of Whiskey in order to settle his nerves following the encounter, but insisted he was not drunk.

Sheriff Fred Diamond, after listening to their basic story, first put Hickson and Parker in separate rooms and questioned each man at length. When the details were compared, both stories not only matched, but corroborated from their specific points of view.

Afterwards, Hickson and Parker were placed in another room that was wired for sound in the belief that if the two men were left alone they would reveal their hoax. Unknown that a tape recorder was running in the interrogation room, the sheriff then left, hoping to catch the men in the middle of a lie. However, the conversation between Hickson and Parker remained true to their story—both of them were genuinely fearful, believing they had really encountered something unearthly. The sheriff after listening to the tape, believed that the two men encountered something beyond the normal.

The local press picked up their tale along with the wire services, and within several days the Pascagoula Encounter was major news all over the country. The Aerial Phenomena Research Organization (APRO), sent University of California engineering professor James Harder to Mississippi to investigate. He was joined by J. Allen Hynek, representing the Air Force. Together they interviewed the witnesses. Harder hypnotized Hickson, but had to terminate the session when Hickson became too frightened to continue.

Hickson and Parker both subsequently passed lie detector tests. Hynek and Harder came to believe the two men's story. And Hynek was later quoted as saying, "There was definitely something here that was not terrestrial."

Although security cameras near the riverbank where the alleged UFO made contact recorded nothing unusual, several eyewitnesses have since come forward who did see something. Among them was retired navy Chief Petty Officer Mike Cataldo – admitting they saw unexplained lights traveling in the skies over Pascagoula the very night of the alleged abduction.

What makes this tale so unusual as well as plausible, was the fact that Hickson and Parker described beings that were not "the Grays." Skeptics always like to tout "the Grays" as always being part of an abduction scenario because it fits so well into the excuse of a media induced hallucination by the amount of attention given to this one type of being. However, these were either androids or perhaps some type of aliens in some sort of biological encounter suits.

Then there is the fact that both men passed not only a comprehension police interrogation, but lie detector tests as well. And lastly there is the matter of the witnesses who later came forward, like Chief Petty Officer Mike Cataldo, who would have done so earlier, but was afraid for his naval career.

On top of that, we have both UFO investigators Hynek and Harder conducting their own investigations and coming to the same conclusion: "There was definitely something here that was not terrestrial." I don't think any skeptics can add anything further to this conversation since Hynek was working for *Project Blue Book* at that time, and if anyone could have discredited the account, it would have been him.

**NOTE:** A few years back a poll was taken and 80% of those individuals asked said that they believed that UFOs were real and that aliens were visiting our Planet.

# Chapter VI: The Government and Related Material

## *Rumor: The Aurora Project – Dark Star*

In the 1990s, the Aurora project rumor was brought to life when it came to the attention of some UFO researchers from documents that they found under the freedom of information act. They stated that the United States government was creating a new type of craft under the direction of aliens. The name "*Aurora*" appeared on a budgeting document with no explanation as to what and where it was being applied. This craft was supposed to have replaced the SR-71 Blackbird. There was a segment on "Unsolved Mysteries" that verified that the name "Aurora" did appear on some budgeting financial sheets. Thus the UFO researchers began calling this the "Aurora Project." **(Two Photographs follow:)**

(From recent photographs: Are they both the Aurora?)

However this name was only used as the financial budgeting project for the real project which was named "Dark Star." This was discovered by an amateur ham radio operator by mistake. On the same "Unsolved Mysteries" segment, this operator in Texas was adjusting his equipment's frequency when he pickup a military broadcast. Moments later he heard a rumbling of thunder in the sky and ran outside. There wasn't a cloud in the sky, however; there was an exhaust trail left by the aircraft which consisted of large puffs of smoke mixed into its trail. He then went back inside his home and when he returned to his radio unit he heard the call sign "Dark Star" being used by the pilot.

This ham radio operator said that the speed of this craft was so fast that it gave multiple breaks in the sound barrier that made it sound like thunder. Rumor of an electromagnetic pulse drive system was developed under Project Squid for this craft. According to Skunk Works (code name for Lockheed – Martin 's Advanced Development Program [ADP] for engineering and technical design of unusual aircraft such as the SR 71 Blackbird) "Dark Star" was an unmanned craft, however; the radio operator clearly heard the pilot give the code name of "Dark Star".

A craft called the X-43 was given to NASA (National Aeronautical Space Administration) by the military some time later, which was believed to be the prototype to the Dark Star project. The improved version is supposed to have wings curved slightly downward with rudders curved slightly inward as in the craft that was found at Roswell. This craft is not only supposed to have radar signature invisibility, but physical invisibility too. The physical invisibility works in such a way that if someone were to view this craft in flight it would appear as if the sky were hazy or slightly out of focus. Supposedly, minutes after this craft has passed any observer, the rumbling of thunder would then be heard. To our knowledge this segment of "Unsolved Mysteries" was never broadcast after its initial airing.

## *Rumor: The TR3B Black Manta - "The Bat"*

Rumors of this *triangular aircraft* have been in circulation since the 1960s. Various studies have been conducted over these years on a special type of anti-gravitational vehicle of flight with this description in mind. Furthermore there are not just photographs taken of this supposedly conceptual vehicle, but recently patents have been taken out on it. This stealth vehicle is the one and only **TR3B Black Manta**.

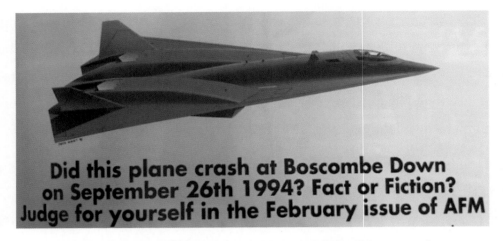

**From the United Kingdom's *Airforces Monthly* magazine**

A prototype (*or perhaps an initial production run*) would seem to have been flying since the late 1980s. Below are perhaps three actual photographs taken of this secretive vehicle of flight. The first one comes from the Belgium UFO flap **(photo below)** between November 1989 and April 1991. It also has been seen on a regular basis around Las Vegas, Nevada from 1990 to 2004. That's pretty close to Area 51 and Nellis AFB.

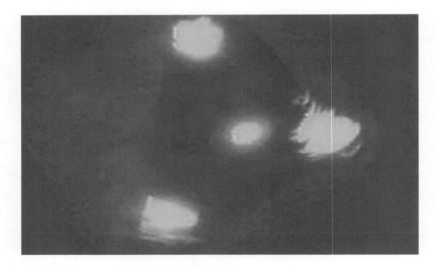

Scientists and engineers have speculated that the propulsion systems incorporated suggests mercury that has been highly pressurized by nuclear energy device which in turn is surrounded by electrical coils to create a plasma ant-gravity field that is governed by a magnetic field to allow the vehicle to maneuver in similar fashion to UFOs at such quantum levels. From *supposedly* eyewitness testimony, the type of noise this aircraft emits is nothing more than a soft hum.

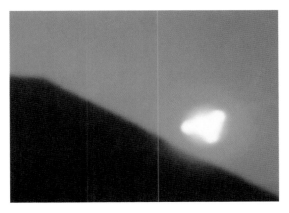

**Photograph taken in Texas, 2012**

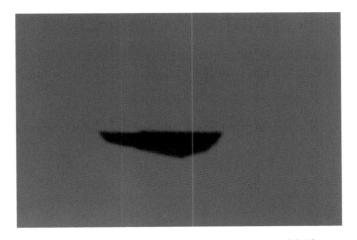

**Photograph taken in Pennsylvania, 2018**

The most stunning part of this story are that patents have been filed recently on behalf of the U. S. Navy. An Aerospace Engineer working for the U. S. Air Force by the name of Salvatore Cezar Pais filed these patents in 2018 for his *Inertia Mass Reduction Device*,...which sure the hell looks very much like the TR3B Black Manta. However, this craft utilizes a resonant chamber (the hull) vibrated by microwaves to create a null space around the craft inciting levitation. This is very similar to the work of John Keely, an engineer of the 19th century that you'll read about in Chapter VII of this book.

## *Fact: The Strange Death of James Forrestal, Secretary of Defense*

James Vincent Forrestal, served as Secretary of Defense from September 17, 1947 until March 28, 1949 and was said to be in favor of telling the truth about aliens and UFOs. In

1949, Truman became angered over Forrestal's continued opposition to his defense economization policies and abruptly asked him to resign. Forrestal who was already under heavy pressure and exhausted from work did not take his dismissal lightly. On March 28, 1949, Forrestal officially left office and a public ceremony was arranged on his behalf.

**Rumor:** James Vincent Forrestal supposedly saw and touched a recovered alien body from possibly the Roswell crash. From the moment after he touched it, he developed severe headaches that continually plagued him for the rest of his short life. This supposedly added to the tremendous pressures he was under as Secretary of Defense that led to his suicide/assassination.

What followed after the ceremony remains mysterious. "*There is something I would like to talk to you about*," William Symington, Secretary of the Air Force, told Forrestal, and accompanied him privately during the ride back to the Pentagon. What Symington said is not known, but Forrestal emerged from the ride deeply upset, even traumatized, upon arrival at his office. Might it had been about Forrestal's insistence on declassifying UFO files?

Forrestal was taken home, but within a day the Air Force flew him to Hobe Sound, Florida, home of Robert Lovett (a future Secretary of Defense). Forrestal's first words were "Bob, they're after me." He met with Dr. William Menninger, of the Menninger Foundation, and a consultant to the Surgeon General of the Army, Captain George N. Raines, chief psychologist at the U.S. Naval Hospital at Bethesda, soon arrived. Robert Lovett later testified that the above words were indeed Forrestal's first words upon meeting him. He thought Forrestal was becoming paranoid for no reason.

For this supposedly reason of suffering a "*nervous breakdown*," it was "*agreed*" that Forrestal needed psychiatric treatment and that the best place was Bethesda Naval Hospital.

**Question:** *However, Forrestal was then a civilian at the time, so why was he taken to a military only facility and on who's direct authority?*

**NOTE:** In 1984, Dr. Robert P. Nenno, a young assistant to Dr. Raines from 1952 to 1959, disclosed that Raines had been instructed by *"the people downtown"* to put Forrestal in the VIP suite on the sixteenth floor of the hospital. Dr. Nenno emphasized that Raines's disclosure to him was entirely ethical, but that "he did speak to me because we were close friends."

Upon arrival at Bethesda Naval Hospital James Forrestal declared he didn't expect to leave the hospital alive. He was proven right seven weeks later.

After weeks of treatment, all five doctors assigned to him agreed that James Forrestal was making a good recovery and would be out of the hospital soon. His security measures were loosened and he was even permitted to walk on the whole 16th floor. The only room that had secured windows was his own. He could have committed suicide at any time by jumping out of the window, but did not. He was allowed to shave with a razorblade, but didn't slit his wrists. His room had blinds from which he could easily have hung himself at any time. He did not.

*Only Dr. Raines claimed that Forrestal was very suicidal and could commit suicide at any time*, yet the other four doctors did not share that opinion. As a matter of fact, they openly disagreed with Dr. Raines.

**Question:** *Both Dr. Raines and Dr. Nardini testified under oath that Forrestal made no attempt at suicide while under either's care. So where did Dr. Raines get the notion Forrestal was suicidal? Dr. Raines did state Forrestal mentioned suicide on many occasions to him privately, but strangely he was the only one of all the doctors to testify to this.*

More peculiar, Dr. Raines' second in command (Dr. Smith) gave testimony that indicated that not much was wrong with Forrestal. In direct contradiction to Dr. Raines (and remembering that Dr. Raines was handpicked by the surgeon general to "care" for Forrestal), Dr. Smith stated this:

"He was a man who not only was mentally alert, but continued to maintain an active interest in all current matters on a level compatible with his broad public service and wide experience. These conversations ran a gamut from a discussion of matters of purely local interest to various philosophies and ruminations that touched on the behavior patterns of all people under various circumstances of stress and his astuteness and acumen were such that his comments and discourses were pregnant with comprehensive significance. I was more often the listener then the speaker."

On May 21st, the last day of his life, he seemed in the best of health. All the people who saw him that day confirmed this before the Willcutts Committee. This was to be his last day in the hospital as his brother called in stating he would pick James up on May 22nd. Only hours before his brother arrived, James Forrestal was dead by supposedly jumping out of a 16-story window that night.

Another interesting fact regarding Forrestal's normal Naval watch, a U.S. Navy corpsman Edward Prise was relieved of his duty that night by a new man, Robert Wayne Harrison, Jr., who had been assigned as a replacement. He never met Forrestal and had only started serving at Bethesda one week before James Forrestal arrived.

According to Harrison, Jr., he checked on Forrestal at 01:45 and he was apparently sleeping. At some point into his shift, Harrison, Jr. left the room to run an errand. When Harrison, Jr. got back to the room, he was shocked to the core to see that Forrestal was not in his bed and the room's windows were open. Harrison's Jr. then raced to the window: the cord of Forrestal's dressing-gown was tied to the radiator near the window. Clearly, there was an attempt at death by strangulation. It turned out, however, that Forrestal's weight caused the cord to snap and Forrestal fell ten floors to his death; something that absolutely no one could have had a chance of surviving. Oddly though, there's also the matter of why precisely Robert Wayne Harrison, Jr. left the room. His official orders, as was Prise's, **Forrestal was never to be left alone!**

**Question:** *What was the errand Robert Wayne Harrison, Jr. ran and for whom? Since this was against his standing orders, unless it was an officer who had the authority to counter-command those orders, Robert Wayne Harrison, Jr. should have been charged at the least with <u>Dereliction of Duty</u> and <u>Desertion</u> of his post. However, he wasn't charged at all. Why not? Having worked in secured facilities in the military, if I had done exactly what Robert Wayne Harrison, Jr. had done, I would have spent the rest of my life in the military prison at Fort Leavenworth, Kansas.*

The story gets even more suspicious at this point. The crime scene photographers were not permitted to enter his room until it was *"cleaned."*

**NOTE:** A page from the Declassified CIA MANUAL: A Study Of Assassination (1953). Please remember James Vincent Forrestal died on May 22nd, 1949.

> assassination. A hammer, axe, wrench, screw driver, fire poker, kitchen knife, lamp stand, or anything hard, heavy and handy will suffice. A length of rope or wire or a belt will do if the assassin is strong and agile. All such improvised weapons have the important advantage of availability and apparent innocence. The obviously lethal machine gun failed to kill Trotsky where an item of sporting goods succeeded.
>
> In all safe cases where the assassin may be subject to search, either before or after the act, specialized weapons should not be used. Even in the lost case, the assassin may accidentally be searched before the act and should not carry an incriminating device if any sort of lethal weapon can be improvised at or near the site. If the assassin normally carries weapons because of the nature of his job, it may still be desirable to improvise and implement at the scene to avoid disclosure of his identity.
>
> **2. Accidents.**
> For secret assassination, either simple or chase, the contrived accident is the most effective technique. When successfully executed, it causes little excitement and is only casually investigated.
>
> The most efficient accident, in simple assassination, is a fall of 75 feet or more onto a hard surface. Elevator shafts, stair wells, unscreened windows and bridges will serve. Bridge falls into water are not reliable. In simple cases a private meeting with the subject may be arranged at a properly-cased location. The act may be executed by sudden, vigorous [*excised*] of the ankles, tipping the subject over the edge. If the assassin immediately sets up an outcry, playing the "horrified witness", no alibi or surreptitious withdrawal is necessary. In chase cases it will usually be necessary to stun or drug the subject before dropping him. Care is required to ensure that no wound or condition not attributable to the fall is discernible after death.
>
> Falls into the sea or swiftly flowing rivers may suffice if the subject cannot swim. It will be more reliable if the assassin can arrange to attempt rescue, as he can thus be sure of the subject's death and at the same time establish a workable alibi.
>
> If the subject's personal habits make it feasible, alcohol may be used [*2 words excised*] to prepare him for a contrived accident of any kind.
>
> Falls before trains or subway cars are usually effective, but require exact timing and can seldom be free from unexpected observation.
>
> Automobile accidents are a less satisfactory means of assassination. If the

The official Navy review board, which completed hearings on May 31, waited until October 11, 1949 to release only a brief summary of its findings. The findings stated only that Forrestal had died from his fall from the window. However, it did not say what might have caused the fall, nor did it make any mention of a bathrobe sash cord that had first been reported as tied around his neck. The full report was not released by the

Department of the Navy until April 2004 in response to a Freedom of Information Act request by researcher David Martin.

The Navy's official autopsy report has never been released, as also were the photos of the body. They all conveniently "*went missing.*" The cord around his neck was never mentioned in the media at the time.

There were a few other discrepancies as well. These include: no damage to his cervical vertebrae was found, which rules out hanging; the sash was found intact around his neck so it was never attached it to a fixed point like the radiator below the window. Another discovery were scratch marks outside the window as though James Forrestal struggle to stay inside the building.

The official report could not establish what exactly had happened to him, leaving it at that. The official conclusion, James Forrestal fell from his window and it was not the fault of any personnel at Bethesda.

However, there were a few quirky things we must mention here. First, the members of the Willcutts Committee Board were a group of Navy doctors; they were not police investigators, nor even military police. Second, when the nurse Dorothy Turner described the deserted bed she saw with Forrestal's bedclothes half turned back, the board at that time had in its possession the photographs taken by hospital corpsman McClain of a bed with a bare mattress, but none of them asked anyone about the contradiction. *Very strange!*

## *Rumor: The Black Pyramid*

There is a mysterious location in Alaska known as the Black Pyramid with "No Trespassing" signs scattered in great quantities about, which also appears to be under military control. The Black Pyramid Theory first emerged back on May 22nd, 1992, when China tested out their one-megaton device which resulted in incredible seismic patterns being spotted underneath the Earth's crust. This gave the experts a reason to scan out the Earth which is when they also came across a very large pyramid underneath Alaska.

According to Doug Mutschler, "a retired U.S. Army Counterintelligence Warrant Officer," geologists had used the detonation to undertake a seismographic study of the earth's crust, only to find "a pyramid structure larger than Cheops" underground, somewhere west of Mt. McKinley.

However, no buildings are there and the area is barb-wired fenced off, which is protecting absolutely nothing as seen by the naked eye. There also are spots in the ground where ground where nothing grows and snow stays melted. Rumor states that there is a Black Pyramid underground and is a source for various electro-magnetic disturbances that do occur in the region. UFOs also have been seen frequently in this area and are said to have an underground base there.

**NOTE:** Alaska has the most recorded sightings of UFOs than any other state.

## Investigating Agencies of the Government on UFOs

**AATIP = Aerospace Advance Threat Intelligence Program (Part of the CIA)**
**AI = Army Intelligence**
**CIA = Central Intelligence Agency**
**CID = Criminal Investigation Division (United States Army)**
**DARPA = Defense Advance Research Project Agency**
**DIA = Defense Intelligence Agency**
**FBI = Federal Bureau of Investigation**
**NI = Naval Intelligence**
**OSI = Office of Special Investigation (United States Air Force)**
**OSI = Office of Special Investigation, Special Counter-Intelligence Unit (United States Air Force - "Men in Black")**

## Types of UFOs Sighted

According to the numerous amount of sightings of UFOs there are several types that have been seen. These are both large and small. These types are as follows:

**Sphere or Orbs** - These are most likely small and unmanned and AI drone controlled. They are possibly use to collect data on our atmosphere, bacteria, soil, etc.

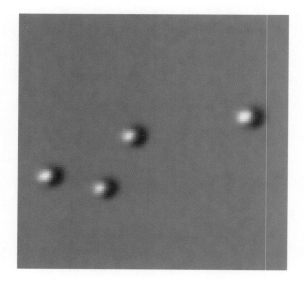

**Dome or Saucer** - These are probably used as scout ships with no more than 3 o 4 Occupants

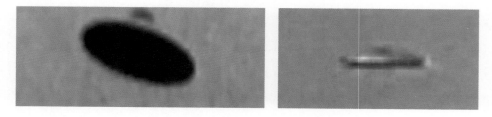

**Cylinders, Cigars or Tick Tacks** - These are medium size craft with a greater Occupancy.

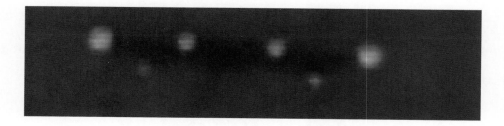

**Delta shaped** - These are also a medium size craft.

**V Shaped or Triangle -** These are large craft or mother ships used to hold the smaller Craft and circumventing time-space continuum.

**Planetary Shaped Mother-ship -** As encountered by **JAL Flight 1628** - By the description by the pilots, Saturn shaped vehicle large enough to at least fit four Boeing 747s into it.

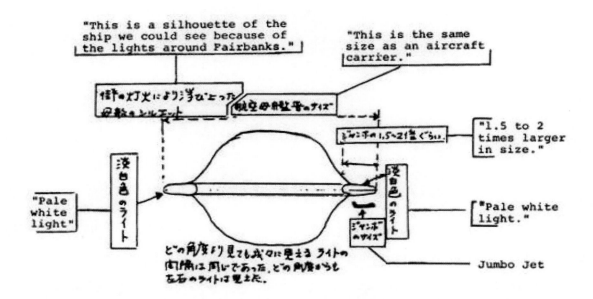

## Some Criteria Noted

Some criteria noted is their ability to maneuver in any direction instantly. This defies the laws of gravity. If one of our aircraft were able to perform these types of maneuvers, the pilot would be crushed by the G Force of our planet. This means that these UFOs have an inertia canceller. In other words, these UFOs have created a gravity system within their vehicles which counters our gravity, thus stops the G Force of our planet from affecting their occupants.

**Rumor:** It is said that a piece of one of their Delta shaped craft was analyzed and was found to be 100% aluminum, which cannot be made by *our-state-of-art technology* as it needs to be made in an environment that lacks any traces of oxygen. It is believe that this piece was part of a magnetic drive system, which allows these vehicles to move in any direction instantly. Nuclear explosions can affect electromagnetic fields, which maybe the reason we are seeing more of these craft because of our nuclear testing, which can be monitored from space. These vehicles may use antiprotons or antimatter (Zero Black Energy) to power there electromagnetic drive system.

It was recently learned that UFOs can control heat signature and may not be picked up by flair cameras...if the occupants of the craft so desire, similar as with radar. However they do admit a low frequency hum that can be picked up by a parabolic microphone.

## Fact? Bob Lazar

**NOTE**: Robert Lazar is a controversial figure within the UFO community. There are those who firmly believe in him, while other respected investigators (like the late Stanton Friedman) absolutely do not.

In order to determine Lazar's authenticity, background checks were performed by many people. However, the results often proved to be ambiguous and inconclusive at best. MIT and Caltech, which he claimed he had attended, had absolutely no record of him nor did Los Alamos National Laboratory where he claimed he had worked. A 1982 lab phone book did list his name, but according to the phone book designation, he worked for a vendor—not the lab itself.

However, in the same token, we have John Lear, now retired from flying, who vouches for Lazar's story and videotapes of the US Air Force's night testing of UFOs. John Lear has flown both commercially and in missions worldwide for the CIA. He has amassed over 19,000 hours of flight time, and has flown in over 100 different types of airplanes and is the only pilot to hold every airman certificate issued by the FAA, and still holds many world records.

There has also testimony from Chase Brandon, a 35-year CIA veteran, along with other former military and CIA operatives who had visited Area 51, and have come forward and verified the same information. Therefore, with fairness in mind as well as seeking the truth, we have presented Lazar's accounts because of such overlapping details that give him plausibility of high moral character.

Also, we, the authors must add something else at this point. It is from talking with other people who have had previous involvement or experiences with the United States government in the past that we must conclude that anything is possible. The loss of careers, the altering of backgrounds, altering of military files, tampering with trial evidence or given statements along with accusations of unjustly being sentenced to prison have all been reported.

Moreover, one does have to remember how the good people of Roswell were threatened and treated by military officers and the FBI in complete disregard of the federal laws and oath of office they are suppose to uphold as well as the very Constitution of our republic. It is too bad that such idealistic notions only apply to people of conscious and, and not to those who lack any moral integrity and excuse criminal behavior in the name of *"National Security."*

With this as a backdrop, one must make up one's own mind about Bob Lazar…and the tales you are about to be told.

Robert Lazar in the 1990s went public on a television special entitled "Alien Secrets – Area 51." Bob Lazar made a claimed that he was working on reverse engineering of an alien space craft propulsion system at a location known as Sierra 4 or S-4 in Area 51. Many UFO researchers believe that he was referring to Groom Lake, however in the description given by Bob Lazar, the hanger was not located in Groom Lake, but instead at Papoose Lake.

Groom Lake is supposedly where our latest technology is tested and a security clearance for one area does not necessarily give clearance for another area of the facility. Groom Lake clearance maybe "Top Secret," but alien technology is most likely considered "Above Top Secret." Papoose Lake is rumored to be where reverse alien technology is taking place. This location is 15 miles south east of Groom Lake.

According to Bob Lazar, he worked at S-4 from 1988 to 1989. The hangers at this location are built into the side of a mountain and the exterior of the hangers are designed to look exactly like the surface of the mountain in this dry lakebed. This was done so that any satellite passing overhead would not detect the location of the hangers.

It is also rumored that at this location there is a surface level with four other levels underground, of which the lowest level is categorized for cryogenic storage and laboratory for deceased alien beings.

The workers are flown from McCarhan Airport, part of Nellis Air Force Base, to Groom Lake via Janet Airlines (possible a synonym for Joint Army Navy Extraterrestrial Transport), which is owned and operated by the government. Once arriving at Groom Lake, the individuals who are working on extraterrestrial technology are then bused to Papoose Lake. On one occasion Bob Lazar was brought to hanger 5, which is one of 9 hangers to the right in this complex. There upon walking in, he observed an alien craft that was in perfect condition. At this point he thought that these flying saucers were being made by the United States Government. However, he soon realized that this was not the case. He noted that there were 9 alien space craft all total and 6 of them appeared to be damaged. Bob Lazar also said that the government had extracted element 115 Moscovium from the alien craft. He was criticize by scientists at that time saying there was no such element. In 2004, element 115 Moscovium was discovered by some of our scientists along with element119 Ununennium. This fact also adds creditably to Bob Lazar's story.

In the course of performing his work assignment, he had to go into the craft to view the placement of the propulsion system. He discovered upon entering the craft that the interior was not designed for a normal size human to utilize. Everything inside was of a smaller scale. He found in the lower level of the craft, the object he was sent to work on: *a gravity amplifier*. Upon coming up to the main level, the walls of the ship became transparent and he was able to view the inside of the hanger. Going outside the craft he found that the original shaped remained the same and that the inner walls of this craft became a 360-degree viewing screen. He stated, "that inside the craft everything seemed to be molded with no edges and appeared unearthly."

Bob Lazar stated that he had worked in Area 51 for 4 months. Then one night he was having a conversation with his closest friend Gene Huff. The subject of what Bob Lazar did for the government arose and Bob told his friend the true nature of this work. Shortly thereafter, around 21:00 hours at night on March of 1989, Bob decided to show his friend a test flight and took him to an area outside the base (believed to be on Route 375 known as The Extraterrestrial Highway in Nevada). Here they could view a craft as it went through its maneuvers.

On schedule, a UFO appeared. They watched this UFO hover then get brighter as it came closer and once more went back over the mountain range. According to Bob, they went back to this location several times and on their third visit they were apprehended and taken in for interrogation. At this point, **the government agents threatened to kill Bob and his wife**, but soon afterwards, released them. However, government agents ransacked Bob's house and tapped his phone. According to Bob Lazar, the government erased his identification and schooling records and made it appear that he never existed in that he could not even get a copy of his birth certificate.

Robert Cechsler (NASA Mission Specialist from 1974 to 1978) heard that the United States government had an alien vehicle and investigated Bob Lazar's credentials to see if Bob really worked in Area 51. Obtaining a copy of Bob's W-2 form, it revealed that Bob Lazar was being paid by United States Naval Intelligence. In checking with the Internal Revenue Service, they stated that this was a genuine document. The identification numbers were highly classified, and because of this, it proved impossible to track down Bob's original employer that was associated with this W-2 form.

George Knapp reporter for KIAS – TV in Las Vegas also decided to check on Bob Lazar story and was able to track down a corporation that had Bob Lazar listed in their company directory. He spoke with Los Alamos Laboratories to see if Bob worked for them. However, they refused to cooperate, saying that Bob was not located in their files. When George showed Bob's name in their company directory, they said well maybe he worked there, but for a different company. Astronaut Dr. Edgar Mitchell of the Apollo 14 mission and the 6$^{th}$ man to walk on the moon spoke with Bob Lazar *and believed most of Bob's story*. He further stated that Roswell was a cover up and that it was a true occurrence. In 2015, Dr. Robert Krugle confirmed that Bob Lazar did work for Los Alamos National Lavatory.

## Fact: The Black Knight

What has been floating around the Earth (for perhaps thousands of years), is painted black, can change its trajectory at will, and transmits Long Delay Echoes while making NASA the butt of more of our jokes? Why the Black Knight Satellite of course!

The first picture of the Black Knight was snapped back in 1960. Since then, *for something that is supposedly to be a mere fairytale or legend at best*, there has been a plethora more photographs of the *Black Knight* taken, released and now in public hands, many courtesy of NASA, itself, and its astronauts. Yet, NASA refuses to concede its very existence even though it filmed it, and has recently gone to length as blaming any photographs of the object on a small thermal blanket from the International Space Station lost in the sea of space in 2008 floating around the orbit of the Earth. This flies in the face of logic, reason, or any rational thought! **(Photograph following)**

**The Black Knight as photographed in 1967 by NASA.**

**Question:** *How do you suppose that a space blanket lost in 2008 managed to travel back in time to the 1960s? Was it transported by the same "magical" way chimpanzees and test dummies of Project Mogul traveled back in time to Roswell in 1947 from the 1950s?*

## Early History of the Black Knight

The Black Night UFO is an enigma wrapped in a puzzle. This satellite has been described as very ancient and possibly circling our planet for 13,000 years. Furthermore, it is said to be transmitting a coded radio frequency that can be monitored by ham radio. And these coded radio frequencies have been received by people such as Tesla and Marconi no less.

**NOTE:** It was actually Nicola Tesla who was the first to pick up unknown repeating radio signals from space in 1899, and later, *publicly announced it at a conference.*

An attempt was made eventually to establish 2-way communications with the satellite in 1928. A Norwegian scientist by the name of Eindhoven Holland transmitted a series of radio waves in rapid succession while aboard a French research vessel in the South China Sea. He then examined the return echoes and noticed they were not as neatly placed as they were sent, ranging from 1 to 30 seconds in delay time. These discrepancies at the time were categorized as being caused by magnetic disturbances within the Earth's atmosphere.

In 1953 (four years before the U.S.S.R. launched Sputnik I), this unknown satellite was sighted by Dr Lincoln La Paz of the University of New Mexico. And more reports of similar sightings trickled in from around the world about the mysterious space vehicle. A then panicked U.S. Department of Defense hurriedly appointed distinguished astronomer Clyde W. Tombaugh (discover of Pluto) to run a search on the mystery object to determine what it was and possibly what type of threat it possessed. It was at this point the mysterious object became known as…*the Black Knight!*

In 1954, newspapers (such as the St. Louis Post Dispatch and the San Francisco Examiner) reported an announcement made by the US Air Force that *two satellites* were found to be orbiting the Earth. One has to remember that this was still at a time when no nation yet had the ability to launch such space vehicles. The existence of the Black Knight had been established by numerous pieces of evidence (including the reception of those Long Delay Echoes or LDEs for short), and was confirmed by the US Air Force.

However, these short wave patterns in 1970 were thoroughly analyzed by astronomer Duncan Lunan, who came to a startling conclusion. Lunan said he had deciphered the transmissions by plotting a vertical axis of the transmitted pulse sequence with a horizontal axis of the echo delay time. And in doing so, not only revealed its origin as the *Epsilon Bootes* star system, but that is was most definitely of extraterrestrial origin. The star chart he produced showed the star system *as it would have precisely looked like 13,000 years ago!* Maybe he should have gotten together with Betty Hill and compared star charts. *Epsilon Bootes is an A 2 star system which is younger than our own and most likely unable to have an advance race. (See page 8 Goldilocks Zone in this manuscript)*

## The First Photographs of the Black Knight

In 1957, Dr. Luis Corralos of the Communications Ministry in Venezuela photographed it while taking pictures of Sputnik II as it passed over Caracas. The strange vehicle was unlike of that of Sputnik I and II. The Black Knight Satellite orbited Earth from east-to-west. Sputnik I and II, on the other hand, orbited west-to-east using Earth's natural rotation to maintain orbit.

On September 3, 1960; seven months after the satellite was first detected by radar, a tracking camera at Grumman Aircraft Corporation's Long Island factory took a photograph of it. People on the ground had been occasionally seeing it for about two weeks at that point. Viewers would make it out as a red glowing object moving in an east-to-west orbit. Most satellites of the time, according to what little material I've been able to find on the black knight satellite, moved from west-to-east. Its speed was also about three times faster than normal. A committee was formed to examine it, but nothing more was ever made public.

However, in March of 1960, Time Magazine published a story on The Black Knight Satellite. In this article, the Department of Defense claims the satellite to be a space derelict from a previous mission. And to quote Time Magazine:

"The Department of Defense proudly announced that the satellite had been identified. It was a space derelict, the remains of an Air Force Discoverer Satellite that had gone astray. The dark satellite was the first object to demonstrate the effectiveness of the U.S.'s new watch on space. And the three-week time lag in identification was proof that the system still lacks full coordination and that some bugs still have to be ironed out."

After which the story faded quickly out of public memory and was all but completely forgotten.

Below is a photograph of the Air Force Discoverer Satellite. Why it is so obvious! You can clearly see the near perfect resemblance between the two vehicles (*with sarcasm and lots of laughter*)! **(Photograph following)**

**(A Photograph of the Air Force Discoverer Satellite)**

Again, how can anyone in their right mind say that the Air Force Discoverer Satellite is in reality the Black Knight? I don't mind the lie as half as much as NASA's opinion of us as being so stupid and gullible to believe that these two different space vehicles are one and the same. Let's be real, here! Even the color doesn't match! And then is the little annoying fact of why so many notable people such as Nicola Tesla tracking it and/or photographing it...*since 1899!* Was the Air Force Discoverer Satellite place in orbit back then? What about the space blanket lost in 2008?

There is also a story about Astronaut Gordon Cooper, who sighted this ancient satellite on a multiple orbital mission. On his very last orbit, he described to the Muchea tracking station, in Australia (also able to pick it up on radar traveling in an east-to-west orbit), a green glowing object ahead of him in space and it seemed to be closing in on his capsule. The news media immediately picked up on the story. However, after his landing, the reporters were not allowed to ask Gordon any questions. The official statement was that Gordon was hallucinating because Air Force and NASA officials said he was breathing a buildup of carbon dioxide in the capsule. I guess he wasn't the only one inhaling the stuff!

But the sightings didn't stop there. One of the more famous photos floating around the internet and old UFO magazines is from STS-088, as spied from *Endeavor* back in 1987. One has to remember that all this was before the advent of computer imaging…and therefore, a little harder to fake.

## *Rumor*:  The Moon Monolith

In 1969, the Apollo 10 astronauts Stafford, Cernan and Young supposedly were the first to film an extraterrestrial space beacon, dubbed *the Monolith*. It was somewhat like, but smaller than, the one in Arthur C. Clarke's book/movie "2001". (It is said this is where he got the idea from.) They were not the first astronauts, however, to spot this space beacon.

As the story goes, Apollo 10 went out of its way to film the structure from every angle. The Monolith acted like a communication beacon. It also supposedly had a message on it, in addition to a map of the extraterrestrial civilization, which placed it there and how to get to them." (The Apollo 10 astronauts are said to have brought back the message captured on film.)

The Monolith's energy field, however, affected Apollo 10's instrumentation so badly that the capsule almost didn't make it back. And as soon as the film Apollo 10 took came back to Earth, it was immediately confiscated and whisked off to never be seen again. But this tale in particular is not verifiable.

## *Fact:*  The Martian Monolith

However, there is more to the *Monolith* story than written above…and it comes from Buzz Aldrin, first man to walk the face of the moon. In a C-SPAN interview, the late astronaut told of a monolith on the Martian moon, Phobos. And there are also pictures. The photograph below was captured on a *Mars Surveyor* telescopic image of the side of Phobos visible from Mars. **(Photograph following)**

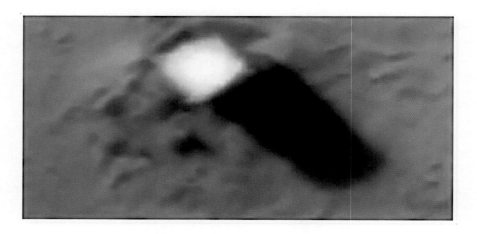

**The Monolith on the Martian moon, Phobos**

By one's own eyes, one can see that this is a rectangular obelisk. With no erosion present at all, since when does nature make rectangular rocks and slam the precisely into small moons? It is possible, but not highly probable. And this slab looks more like it's been planted rather than have impacted, although I can give you reasons for both. But by the very angle of 90 degrees it looks to standing at in the ground is most puzzling.

## *Fact: Astronauts coming forward about UFOs*

Over the years, there have been a number of American astronauts, and Russian cosmonauts who have come forward and talked about their UFO experiences. On the American side for example, Edgar Mitchell stated that he knows that we are not alone in the universe, that there is intelligent alien life, that the United States government knows it, has known it for decades, and has concealed the knowledge from the general public.

In further interviews, Mitchell confirmed the existence of Majestic 12 and the ongoing cover-up by the United States government on UFOs. He also has confirmed there has been direct contact between the United States government and aliens. He then went on further to admit that this was all made possible by the National Security Act of 1947.

But Edgar Mitchell is not alone among astronauts that have admitted to seeing UFOs. Buzz Aldrin admitted from 2005 to seeing a ringed object keeping pace with the Apollo 11 flight. By Aldrin's own account, film exists of this object, taken by Aldrin himself, *which was confiscated by the CIA.*

Then there was Dr Brian O'Leary who passed away on July 28th, 2011 shortly after making a statement about UFOs existing and their types of propulsion systems.

Neil Armstrong during a rare interview made this statement: "It was incredible, of course we had always known there was a possibility, the fact is, we were warned off (by the Aliens)! There was never any question then of a space station or a moon city."

Astronaut Gordon Cooper is another, who told a United Nations panel in 1985 that he had witnessed UFOs personally. Cooper was one of the original 7 Mercury astronauts. He said he hoped that the U.N. would pursue their study of UFOs and extraterrestrials in a responsible manner, so more professional people who have actually seen UFOs and have knowledge of other-worldly life would feel safe to come forward. Over the years, he has been both candid and persistent in his testimony.

The accounts of these men cannot be excused away or dismissed. They were trained and trusted observers who have given faithful witness of their experiences and the experiences of others. And since these details are so readily available to the reader elsewhere, we, the authors, feel there is no reason to go on any further in detailing their statements, which would take a considerable amount of time and space to categorize. However we did include several accounts by several Russian cosmonauts.

**NOTE:** *NASA now stipulates in its agreement with astronauts that **they are never to talk about UFOs** to the public. What does that say?*

## Account By COSMONAUT GENNADIJ STREKHALOV:
### (MIR Mission 1990)

"On the last two flights I saw something. During the flight of 1990, I called Gennadij Manakov, our commander: "Come to the porthole". Unfortunately, but this is typical, we did not manage to put a film in the camera quickly enough to film it. We looked on Newfoundland. The atmosphere was completely clear. And suddenly a kind of sphere appeared. I want to compare it with a Christmas tree decoration, beautiful, shiny, glittering. I saw it for ten seconds. The sphere appeared in the same way as it disappeared again. What it was, what size it had, I don't know. There was nothing I could compare it with. I was like struck by lightning by this phenomenon. It was a perfect sphere, glittering like a Christmas tree decoration. I reported to the Mission Control Center, but I did not say that I have seen a UFO. I said I saw a kind of unusual phenomenon. I had to be careful with the choice of my words. I don't want someone to speculate too much or quote me wrong."

## Account By COSMONAUT GENERAL PAVEL ROMANOWICH POPOVICH: (Soviet Air Force)

"I had only one personal encounter with something Unknown, something we could not explain. It was in 1978, when we flew from Washington to Moscow. We flew in an altitude of 30,600 feet. And suddenly, when I looked through the windshield I noticed something flying about 4500 feet above us on a parallel course - a glowing white equilateral triangle, resembling a sail. Since our speed was 600 mph, the triangle must have had a speed of at least 900 mph, since it overtook us. I called the attention of all passengers and crew-members on it. We tried to find out what it was, but all attempts to identify it as something known ultimately failed. This object looked like a UFO and it remained unidentified. It did not look like an airplane, since it was a perfect triangle. No airplane at that time had such a shape."

## Account By COSMONAUT MAJOR GENERAL VLADIMIR KOVALYONOK: (Saljut VI Mission 1981)

"Many cosmonauts have seen phenomena which are far beyond the experiences of earthmen. For ten years I never spoke on such things. The encounter you asked me about happened on May 5, 1981, at about 6 PM, during the Saljut Mission. At that time we were over the area of South Africa, moving towards the area of the Indian ocean. I just made some gymnastic exercises, when I saw in front of me, through a porthole, an object which I could not explain. It is impossible to determine distances in Space. A small object can appear large and far away and the other way around. Sometimes a cloud of dust appears like a large object. Anyway, I saw this object and then something happened I could not explain, something impossible according to the laws of Physics. The object had this shape, elliptical, and flew with us. From a frontal view it looked like it would rotate in flight direction.

"It only flew straight, but then a kind of explosion happened, very beautiful to watch, of golden light. This was the first part. Then, one or two seconds later, a second explosion followed somewhere else and two spheres appeared, golden and very beautiful.

"After this explosion I just saw white smoke, then a cloud-like sphere. Before we entered the darkness, we flew through the terminator, the twilight-zone between day and night. We flew eastwards, and when we entered the darkness of the Earth shadow, I could not see them any longer. The two spheres never returned."

## *Multiple Confirmations of Alien Technology*

We have been inundated with a flood of new technology, some of it rumored to have been introduced by either aliens or reversed engineered. Some of these creations and tales have more than a plausible UFO link. We've taken the time to list a few of the more significant pieces of hardware, which may be evidence of our government *"making a pact with the devil"* or in the least—becoming more efficient in reverse engineering. It is something to be aware of…and very concerned about. And the reason for this…our very individual lives may be at stake.

## *Fact: The Piney Woods Incident, Cash-Landrum*

The Cash-Landrum case is about as good as it gets. As previous stated, both women (and in a lesser degree, Vickie's Landrum's grandson) contracted the unexplainable illness of radiation poisoning while witnessing a UFO being test piloted by United States military personnel (as evident from the helicopters that escorted it and the admission of a pilot who flew the mission). But this was not the end of the story.

An after-the-fact revelation occurred on the television documentary "UFO Cover-Up Live" where two US intelligence agents that worked in Area 51 explained to the two women (*brought onto the documentary show*) what really had occurred that night.

The two supposed government agents informed the women that the craft, which have been seen, was given to the United States government by aliens from Zeta 1 Reticuli as part of an exchange program. They further stated that the craft was not designed for human pilots and that there were US pilots controlling the vehicle. The US pilots were having difficulty with the craft because they were not familiar with all the controls and radioed the base that they were going to crash. That was why the helicopters were sent out. Both Betty Cash and Vicky Landrum died some years later of cancer, which was caused by the radiation emanating from the vehicle.

The experimental military craft theory was further supported by Dr. McClelland, Betty's original physician (*interviewed in 2008 by the UFO Hunters*), who said there were people "in the know" had stepped forward and told him that there was an accident with an aircraft called the "WASP 2". However, the US Army WASP 1 was nothing more than a VTOL aircraft (it looked like a flying can) and bears no resemblance, even remotely, to the same in description as what Cash and Landrum witnessed. However, it does confirm the US was testing an "unknown" vehicle in a civilian populated area!

So what do we have as evidence? We have multiple witnesses of the same craft and fleet of helicopters (including testimony from Oilfield worker Jerry McDonald) on the same evening, we have two women and a male child contracting radiation poisoning at the same time in a civilian populated area (to the contrary of opinions expressed by some Doctors later on, was verified by other doctors at the time, including Dr. McCelland), we have part of a road with its pavement total removed of the sighting area, we have the admission of a pilot from the 160th Air Regiment, who actually flew the mission that particular night, and we have the admission of two purported government agents (*who did have their credentials thoroughly checked*) verifying their backgrounds as well as adding a few more details to the story.

**Bottom Line:** There is little (if any) question the United States government is in possession of alien spacecraft and reverse engineering them. But if absolutely true (as we believe), it does so by endangering the general public through unlawful testing of these vehicles in or around populated civilian areas while refusing to compensate anyone for accidental exposure to their negligence. (This would be like conducting a practice bombing run on the city of Houston.) I can't even begin to tell you how many (both civilian and military) laws the involved US Army personnel and other government participants have broken.

**NOTE:** *According to President Clinton's executive order, anyone injured or killed with regard to these "black" projects are considered **an acceptable loss (collateral damage if you will)**, with the responsible culprits free from any prosecution by EPA and other federal laws while given full immunity! And if you think that is a lie, look up the case of Area 51 workers who are still trying to sue the government for illnesses contracted from the burning hazardous and toxic waste material and chemicals. For confirmation of this see the History Channel movie: The Presidents and UFOs also seen in the Movie: Alien Secrets - Area 51.*

## *Rumor: "Solar Warden"*

This is actually a story that can be considered related to the Cash-Landrum case, however; it involves a man named Gary McKinnon, a Scottish systems administrator *and part-time hacker*. In 2002, he was accused of perpetrating the biggest military computer hack of all time. Looking for evidence of free energy suppression and a cover-up of UFOs, McKinnon (allegedly) hacked into 97 United States military and NASA computers over a 13-month period, between February 2001 and March 2002,

Other than the claim (which we believe to be no doubt true) he deleted critical files from operating systems (which did shut down the United States Army's Military District of Washington network of 2,000 computers for 24 hours), and the deletion of weapons logs at the Earle Naval Weapons Station (rendering its network of 300 computers inoperable and paralyzing munitions supply deliveries for the US Navy's Atlantic Fleet); McKinnon is also accused of copying data, account files and passwords onto his own computer. And this is where the rumor of *"Solar Warden"* begins.

According to McKinnon, part of the data he claimed to have found were photographs, film, and other evidence of alien spacecraft secretly held by various U.S. government agencies. There was also a supposedly high definition picture of a large cigar shaped object over the northern hemisphere, a list of officers' names under the heading of 'Non-Terrestrial Officers' and evidence of the highly compartmentalized program called *"Solar Warden"*; a highly classified and *secret space fleet* that operates in outer space, using antigravity technology. There is also a list of some type of naval ships that don't match any names of any vessels within the United States Navy. The fleet is suppose to consist of 8 Large craft and 43 scout ships or shuttles. Rumor is that the United States Government has a patent for unlimited energy related to time space travel and levitation devices. This is very curious.

**NOTE:** On 16 October 2012; after a series of legal proceedings in Britain, Home Secretary Theresa May withdrew her extradition order to the United States. It is speculated that if a trial had occurred, all this information would have been released (if McKinnon didn't *'vanish'* first) and that's why the extradition order was withdrawn.

Sounds fantastic, doesn't it? Well, here is something more fantastic, supporting evidence that comes from Jan Harzan, the new director of the Mutual UFO Network **(MUFON),** a highly respected UFO organization that takes a genuinely scientific approach to UFO investigation. This organization is staffed by scientists, engineers, and other notable and trained observers. Harzan, himself, received an engineering degree from UCLA in 1993 and has been with IBM for over 35 years.

But more importantly in 1993, Harzan received an invitation from the UCLA engineering alumnus association to attend a talk given by no other than Ben Rich (former director of Lockheed's Skunk Works from 1975 to 1991), at the alumni center. Like Harzan, Rich received his engineering degree from UCLA.

**NOTE**: There is also a story circulating around the Internet about an article supposedly written by Tom Keller and published in the May 2010 issue of the Mufon UFO Journal. In it, Ben Rich had made the statements recorded below along with many outrageous others on his deathbed. Although Tom Keller is a friend of Jan Harzan **and was in attendance** at that very same presentation by Ben Rich, we have yet to find that supposed specific article after a considerable amount of time and effort in trying to locate it. (Tom Keller has written a book entitled: "*The Total Novice's Guide to UFOs*.") We believe more than likely that above mentioned Mufon UFO Jounal article story was a distortion of the actual based alumni lecture.

In the presentation, Ben Rich made the following statements:

- "We found an error in the equations (believed to be referring to Maxwell's) and we know how to travel to the stars."
- "We now have the technology to take ET home."

After the lecture, Jan Harzan approached Ben Rich as he was leaving and asked, "I have a real interest in the propulsion you are talking about that gets us to the stars. Can you tell me how it works?" Ben Rich then stopped and looked at Harzan and asked him if he knew how ESP worked. Jan says he was taken aback by the question and responded, "I don't know, all points in space and time are connected?" Rich lastly replied, "That's how it works." The former Lockheed director then turned around and walked away.

These statements made in 1993 give credence to reports (such as Cash and Landrum) that the U.S. military has been flying UFOs derived from actual alien craft. And between the two tales (of Cash and Landrum, and Gary McKinnon) and the lecture, there is a good possibility "*Solar Warden*" actually exists.

However, one more piece to this puzzle presents itself. Dr Harold White of NASA is famous for suggesting that faster than light (FTL) travel is possible. Using something known as an Alcubierre drive, named after a Mexican theoretical physicist of the same name, Dr White said it is possible to 'bend' space-time, and cover large distances almost instantly. Announced in January 2012, it is well known that a joint endeavor run by DARPA, NASA, Icarus Interstellar, and the Foundation for Enterprise Development are currently evaluating a number of different technologies, including 'warping' space time to travel great distances in short time frames at faster-than-light speeds. Just maybe it already exists.

## *Fact: Sonic Weapons for Defense & Mass Beachings*

There have been numerous teams of environmental scientists from NOAA (National Oceanic and Atmospheric Administration) and other organizations from around the world that have been investigating mass beaching of whales, dolphins and great white sharks. The first of these mass beaching occurred on the west coast shores of California in 2004.

These environmental scientists responded to these events and found possible links to the United States and other countries' military activities.

What generated this curiosity was when military personnel began retrieving what appeared to be samples of these dead marine animals. Although the scientists at first thought that these were curious incidents, it should have not been a military problem. This eventually provoked the scientists to dig deeper into the matter as more mass beachings occurred. And what they found were internal organs that had been ruptured or otherwise burned, and destroyed.

An initial independent scientific review panel first concluded that the mass stranding of whales and other sea animals, the result of the beachings, were primarily triggered by acoustic stimuli, more specifically, a multi-beam echo-sounder system operated by survey vessels or an advanced type of military sonar. However, this did not explain away the burned or damaged internal organs. Something else had to be the cause of this peculiar anomaly.

**NOTE:** Later on, the United States Navy conceded it was conducting underwater testing of explosive weapons and sonar devices. They have readily admitted that most of these deaths came (and would continue) from detonation of explosives, sonar testing or being hit by ships. This testing already has been responsible for the death of thousands of these sea creatures, but it still didn't explain away all the odd anomalies the scientists found.

The scientists meticulously poured over the data, performing different types of analysis…and they came up with a startling conclusion…the possibility that a powerful ultrasonic device or weapon destroyed the marine creatures' organs.

**NOTE:** Ultrasonic therapy is a method of stimulating the tissue beneath the skin's surface using very high frequency sound waves, between 800 kHz (kilo Hertz) and 2 MHz (Megahertz). Tissue can be heated by the vibration to the point of burning and damaging it if amplitude (power) of the vibration is strong enough.

Again, we have no surprise here. The reason, a quick blurb made by former astronaut Edgar Mitchell, who in a prior recorded interview, told of sonic weapon technology being handed over to the United States by aliens.

The use of sonic weapons underwater has been widely speculated about.

- Ultrasound is used for heating and healing deep body tissue in physical therapy.
- It can disintegrate solids into liquids for industrial use.
- It has long been known that ultrasound in water will kill small water animals.

**NOTE:** The U.S. Navy has confirmed the existence of the **Integrated Anti-Swimmer System,** which uses frequencies between 85 and 100 kHz (Kilo Herz) that are best suited for diver detection surveillance. The ultrasound weapons (IAS) system includes:

- SM2000 (Underwater Surveillance System), a search sonar.
- An underwater warning loudspeaker.
- An underwater shockwave emitter that can force divers to surface, or stun or kill them. However, the US Coast Guard calls it a "nonlethal interdiction acoustic impulse".

In most cases, the detected diver would be warned by the underwater loudspeaker; after that, the shockwave emitter would be fired once as a warning, before it is fired to stun or kill.

**Speculation:** It has been a matter of record that in the past that UFOs have been seen entering, moving under and emerging from bodies of water. When working for the government, there were rumors of sonic weapons that were being developed. These sonic weapons were said to cause the internal organs of an enemy to cavitate (*irregular oscillations*), which led to extreme pain and possible death.

A company named Scientific Applications and Research Association was purportedly to be heading this project. And from their own website:

*"Scientific Applications & Research Associates (SARA), Inc., was formed in 1989 to harness the creativity, innovation and entrepreneur spirit of engineers and scientists. SARA, Inc. is employee-owned and is managed by leaders that each has 20-30 years experience in Defense and Aerospace. SARA is nearly 100 innovative scientists and engineers doing research and development for government, military, and industrial clients."*

However, their involvement (if any) with this or any other specific military project is UNKNOWN.

**Question:** *If this story is true, then you must ask yourself why would the US Navy need such a high-powered weapon where it can fry the insides of whales, dolphins, and sharks?*

**Answer:** It just might prove to be an effective weapon against any hostile UFOs / USOs (Unidentified Submersible Objects) which have been determined to have an electromagnetic force field and propulsion system. Such a weapon would need to be classified above top secret and the government would want all information repressed and confiscated.

**Question:** *Is this a precautionary weapon to be used against alien craft, especially if the US Government has an agreement with, but does not fully trust the aliens of which they have that agreement?*

## Fact: Strange Electromagnetic Fires and Other Things

Around mid January, 2004; in the little Sicilian beach front town of Canneto di Caronia, electrical appliances were bursting suddenly into flames. These unusual if not bizarre phenomena occurred even when these objects were not plugged in! Odd, isn't it? Odder still, besides the hundreds of small fires without apparent cause and in unexpected places (which includes everything from water pipelines, furniture, and to a moving car), there were the spontaneous activation of central locking of dozens of parked cars, cell phones ringing without any calls, and batteries suddenly running down. A navigation system also went up in smoke, electrical and electronic equipment "become crazy", and several standard compasses measured a sudden deviation from magnetic north.

Canneto di Caronia lies in the northern part of Sicily. And all throughout this sleepy town did these fires occurred without any explanation. The local government presumed that these fires were occurring because the power lines were transmitting to much electrical voltage. Thus they cut the power to the town to stop the fires and make the necessary repairs. However, to their surprise, the combustion of fires continued not only without electricity, but also on items such as couches and mattresses as well.

The town's people began panicking and appealed to the clergy, thinking that this was something supernatural and malevolent. The church responded with exotericism rituals to no avail. A frantic Italian government then invaded the town with dozens of technicians equipped with the most sophisticated instruments for measuring electromagnetic fields.

Experts from the Civil Protection, Enel (power company), ARPA (environment protection), the railways company, three telephone companies operating in the area, and two electrostatic and chemistry professors at the University of Messina performed surveys and soil tests. They were all unable to determine root cause to the mysterious disturbances.

A few scientists believed that Mt. Etna may have been the culprit due to a massive body of solid magma intruding from about one mile to about 11 miles deep beneath the mountain. Presumably, this caused a pressurization of the mountain's deep plumbing system, which had fractured the rock. By doing so, trapped methane (a colorless and odorless gas) and possibly other flammable fumes flooded inside the Canneto di Caronia homes, were then sparked to ignition. But there were absolutely no evidence of that happening either. But as quickly as it started, however; it ended, bringing a sigh of relief…if only momentary.

But it was only a short period of relative calm, for the strange curse began to work is wicked Magick again. Hoses connected to water pipes in the bathrooms and kitchens of three different houses, unexplainably began to leak. And to top that off, a vanity mirror placed in one of two small bathrooms burst into fire three times in 35 hours. There was also an entire plantation of eggplant, which turned into thousands of colors as they became seriously damage and began to rot.

But new reports now filtered in about of bright lights in the sky over the sea north of the town. This drew the mayor of the town, Pedro Spinnato, out to the beach to have a look for himself while he captured the new mystery on film. He did observe large orange orbs in the sky at dusk and tried to photograph them, however his face and arms where severely burned while the camera equipment was utterly destroyed by what seemed to be electromagnetic bursts or pulses. It is also curious that these orbs were not sighted until after these fires started to occur.

Finally, the physicist Clarbruno Vedruccio, scientific advisor of Galileo Avionica explained (only after research was carried out for many months) the phenomena may have been caused by a pulsed beam of electromagnetic waves with high potential, traveling above the sea to a height between the surface and 15 meters. This tight electromagnetic beam (transmitted in the band 300 MHz to 3 GHz) would be comparable to the electromagnetic weapons, developed in recent years by the United States Navy, but used quite differently. The UHF (Ultra High Frequency) beam could be used to disable any electronics…or magnetic fields developed by the electronics of any vehicle within that range.

**NOTE:** A highly concentrated electronic beam, similar to a microwave UHF (Ultra High Frequency) beam, can cause such effects as demonstrated by the US Army's latest crowd control weapon that creates a burning sensation of the skin. This type of beam is in a transmitted range of 300 MHZ (Megahertz) to 3 GHZ (Gigahertz). It can also cause the air to ionize, turning the gas into glowing auroras or into orbs such as ball lightning.

When high powered bursts of Radio Frequency (RF) are induced into conductive metal items, heat is produced. It will also induce spikes into electronic circuitry causing malfunctions. However, to penetrate any vehicle's electronic module shielding requires an abnormally powerful RF pulse to cause such a malfunction. And for fires, when a high power RF pulse strikes metal, it induces eddy currents. These circulating eddies of current have inductance and thus induce magnetic fields and can cause repulsion, attraction, propulsion, drag, but particularly…heating effects. To do this, the eddy currents inside metals act like millions of tiny short circuits. As the RF pulse intensity increases, so does the heat it induces, and thus anything around it or it touches can be turned quickly to kindling.

It is no secret that the US Navy's Sixth Fleet frequently conducts training exercises just up the coast from Canneto di Caronia. And rumor has it that the United States Military was present those times and may have had something to do with the unexplainable occurrences. (It was reported at the time that the navy had ships in the waters just to the north of Canneto di Caronia when it all began.)

**Question:** *There seems to be a connection between this story and the United States Navy's underwater sonic weapon. Could they be one and the same?*

**Answer:** Maybe, but I would wager that they are two separate systems that are used jointly against USOs, hitting them above and below the water at the same time.

## *Fact: HAARP (A weapon for Defense against UFOs?)*

The **High Frequency Active Auroral Research Program** (**HAARP**) directs a 3.6 Megawatt signal, in the 2.8–10 MHz region of the HF (High-Frequency) band, into the ionosphere. It does so by either pulsing the signal or transmitting it continuously. Funded by the U.S. Air Force, the U.S. Navy, the University of Alaska, and the Defense Advanced Research Projects Agency (DARPA), they do so for the quaint and stated notion of analyzing effects on the ionosphere for investigation of developing ionospheric enhancement technology for radio communications and surveillance.

As far as we can ascertain the HAARP Project has three Locations, two in Alaska and one in Puerto Rico. Curiously, they are on opposite side of the country covering over 90 percent of our nation and Canada. Can this be so, they can are actively cover most of our country as a defense weapon against UFO's? (See Map Below) Remember the Roswell Crash was during an electrical storm and it was theorized that a bolt of lightning struck one or more UFOs. Could HAARP create an electrical storm in order to down a UFO? It seems plausible.

**NOTE:** Hawaii was not chosen as a location. Could this be due to the fact of its volcanic structure? Could HAARP set off a volcanic eruption in Hawaii to equal or surpass that of Krakatoa?. Krakatoa was a volcanic island in the providence of the Dutch East Indies now known as Indonesia. On august 26, 1883 it set forth a enormous eruption that set 50 foot waves 125 miles killing 36,000 people in its path to Java and Sumatra. Its shattering eruption could be heard 3,000 miles away. It erupted again in 1927 and its inhabitants were evacuated. It is now uninhabited.

There is no doubt that they are conducting some legitimate experiments, which are yielding information for various scientific fields of study. But I also know never to take anything at face value. This is after all a high power, high-frequency directed phased array radio transmitter with a set of 180 antennas transmitting strangely in the Shortwave Radio Band of 2.8–10 MHz region, and by equally strange coincidence just happens to be above the Ultrasonic Frequencies. As Ultrasound is excellent for transmitting through water, Shortwave radio bands has the ability to "propagate" for long distances through the atmosphere, making possible such world-wide communications as international broadcasting and coordination of long-distance shipping.

However, by a press release in 2013, the U.S. Naval Research Lab declared this:

"Using the 3.6-megawatt high-frequency (HF) HAARP transmitter, the plasma clouds, or balls of plasma, are being studied for use as artificial mirrors at altitudes 50 kilometers below the natural ionosphere and are to be used for reflection of HF radar and communications signals. Past attempts to produce electron density enhancements have yielded densities of $4 \times 10^5$ electrons per cubic centimeter ($cm^3$) using HF radio transmissions near the second, third, and fourth harmonics of the electron cyclotron frequency. This frequency near 1.44 MHz is the rate that electrons gyrate around the Earth's magnetic field. The NRL group succeeded in producing artificial plasma clouds with densities exceeding $9 \times 10^5$ electrons cm³ using HAARP transmission at the sixth harmonic of the electron cyclotron frequency (*and increasing the potential for lightning and other ionic disturbances*)."

And by a few of their own findings, they concluded the following as part of their summary:

1) HAARP signals allows VLF remote sensing by the **heated ionosphere** (*which can cause high and low pressure systems to form in the atmosphere*).

2) HAARP signals can enter the magnetosphere **and propagate to the other hemisphere, interacting with Van Allen radiation belt particles** along the way (*and thus can directly affect weather and weather patterns through altering the Earth's magnetic fields and their protection from the solar winds*).

3) HAARP can effect of ionospheric disturbances on GPS satellite signal quality (*disrupting GPS signals along with other types of satellite communications*).

4) HAARP can generate ELF signals (Extremely Low Frequency signals: from 1 Hz to 300 Hz) by heating of the **auroral electrojet** through a modulated signal. (*An electrojet is an electric current which travels around the E region of the Earth's ionosphere.*) But we'll get into this a little later.

**Question:** *With the above stated findings, why would anyone experiment with something so unpredictable that could dangerously affect obviously weather, navigation, and communication signals as well as the planetary environment?*

*And then there is the question of the installations' positioning, which is contradictory to their stated purpose "is to analyze the ionosphere and investigate the potential for developing ionospheric enhancement technology for radio communications and surveillance." Why would they pick the worse possible place in the world to do so? I mean, sitting under the weakest part of the Earth's magnetic field where the solar winds are so predominate in creating enormous magnetic, ionic and meteorological disturbances isn't exactly the smartest thing in the world...unless one wants to capitalize on it...and make things worse. If one was truly investigating such things, I would set up the facility where weather conditions are more normal and stable, like Hawaii.*

*And then there is the final question to ask: why are they using a 3.6 Megawatt signal? They should be able to achieve reflection with a much weaker signal. Understand, the most powerful AM/FM/Shortwave radio stations in the world used somewhere between 50,000 kilowatts to 1 Megawatt of power...by government stipulated regulations and limits. And this is a rarity. Most normal stations are under 50,000 Watts. HARRP is using a signal 72 times as great! So what does this have to do with normal communications then by the power it's using? It doesn't make any sense, now does it?*

**Answer:** What this all comes down to is that there seems to be another purpose than what was otherwise stated *officially*. For example, HAARP Magnetometer readings (as analyzed by experts in various fields of science) indicate that HAARP apparently attempts to take advantage of the energy originating from solar flares, the gravitational pull from the moon and other celestial bodies, *which is seemingly followed by seismic activity along with changing weather patterns and their severity.*

**NOTE:** The HAARP station in Puerto Rico is attached to the Mas Grande Del Mundo Observatory, which held the world's largest reflector dish and where the James Bond movie "*Goldeneye*" was filmed in part is said to have been destroyed by the last major Hurricane that hit the island. It is unknown by this researcher if the HAARP laboratory was affected by this storm as I have not returned to that location.

## *FACT: The Ionosphere*

The ionized (*electrically charged*) part of the Earth's atmosphere is known as the ionosphere. Here, ultraviolet light from the sun collides with atoms in this region knocking electrons loose. This creates ions, or atoms with missing electrons. (This happens to be similar as to how electric current is conducted through wire.) This is what gives the ionosphere its name and it is the free electrons that cause the reflection of High Frequency radio signals (3 MHz to 30 MHz) and the absorption of these radio waves. And this absorption of electronic energy heats the charged particles up...along with a portion of the ionosphere. And that's exactly what HAARP's signal and its harmonics does so well.

Now if we are just talking about ions in terms of radio communications that are prone to atmospheric changes, these experiments would simply cause fading, interference, and noise while bouncing signals around the world. And this is nothing new. It's been going on since before the 1920s! For this reason alone, this type of experimentation would just be a waste of time. And since these affects are so well known, this would mean that communications tale to be nothing more than a cover story to this experimentation. And what would it be a cover story for?

HAARP is a radio telescope in reverse: antennas which send out signals instead of receiving them. HAARP can be also considered a super-powerful radio wave-beam transmitter that heats and lifts areas of the ionosphere by this focused beam. And this, the Navy has already admitted to. This in turn can create winds, atmospheric high and low pressures, sound waves, and generate and even intensify storm systems. It can also interact with the Earth's magnetic field and all that implies.

Think that's impossible? Well, here is a little quote from Secretary of Defense William Cohen in 1997 at the Counterterrorism Conference in Athens, Georgia.

*"Others are engaging even in an eco-type of terrorism whereby they can alter the climate, set off earthquakes, volcanoes remotely through the use of Electro-Magnetic waves. So there are plenty of ingenious minds out there that are at work finding ways in which they can wreak terror upon other nations. It's real..."*

And so was the devastation of Hurricane Katrina! For some strange reason, this minor storm system mysteriously transformed from a tropical depression (*that should had died out*) into a Category 5 Hurricane after passing over Hallandale Beach in Florida! This rapid growth was attributed to the storm's movement over the "unusually warm waters" of the erratic *Loop Current* that somehow penetrated into the Gulf of Mexico from the ocean between Cuba and the Yucatán Peninsula. (*In fairness, this does happen occasionally, but only under rare circumstances.*)

**Question:** *But what does this all have to do with HAARP?*

**Answer:** Glad you asked that question. It would appear by its own magnetometer charts that HAARP was creating the very nightmare of former Secretary of Defense William Cohen's words as stated above.

HAARP's magnetometer chart for the period from Aug 21-28 had a big spike between Aug 24 and Aug 25. This spike occurred right before Katrina made landfall in Florida. HAARP then shut down until October 8$^{th}$, long after the storm had passed and done its dirty work. But there were other weather phenomena afoot at the same time as well. For example, the southwestern United States was baking with temperatures over 100 degrees Fahrenheit while the eastern seaboard was being bombarded by thunderstorms, tornados, and baseball size hail just to mention a few.

**NOTE:** During that downtime from HAARP Alaska, it was surmised that they had either deliberately shut the data off to various weather and academic feeds or somehow loss total transmission. No explanation (to my knowledge) was ever given. You know, it's funny, the very same thing reportedly happens with NASA and its communication feeds whenever there's a UFO encounter around one of its space vehicles. How about that?

And there have been many similar occurrences of disasters that were seemingly timed with the aftereffects of a HAARP experiment. However, what makes this so frustrating, HAARP effects crosses over many boundaries of science, and unless you have an army of scientists at your command to culminate and analyze the information accurately, having definitive proof as to the cause of these disasters becomes extremely elusive. One can only speculate for the most part with what little information one has.

## *FACT: Vibrations, Piezoelectricity, Earthquakes, and HAARP*

The word *piezoelectricity* means electricity resulting from pressure. It is derived from the Greek word *piezo*, meaning to squeeze or press. (The properties of quartz crystals were discovered by the ancient Greeks.) Piezoelectricity is an electric charge (more predominate in certain solid materials such as crystals, certain ceramics, and biological matter than others) given off in response to applied mechanical (physical) stress.

For example, a piezoelectric disk generates a voltage when deformed by changing pressure or exposed to mechanical vibrations. In other words, when a crystal is squeezed with mechanical vibrations, it will transform them into electrical vibrations as alternating voltage. You can think of this as a "microphone" affect: translating sound into electricity, which can be used in the recording of a voice or even music.

It can also perform the reverse function. If you stimulate a quartz crystal with electric pulses, it will translate these electrical pulses into mechanical vibrations. (In other words, the quartz crystal vibrates.) This could then be considered a "speaker" affect. However, this usually takes a special transformer which uses acoustic coupling as part of the circuitry. And quartz crystals perform these and other affects by the way they are cut and shaped.

## Now we come to earthquakes…and HAARP.

**NOTE:** All particles of all matter are made up of vibrating energy as so stated by Quantum Physics. Everything, regardless of what it is, not only vibrates, but has a **resonant frequency**: the predisposition of a piece of matter to oscillate at one particular frequency (greater than any other) with such an intense magnitude that it can literally shake itself apart.

To begin with, earthquakes give off vibrations in the form of seismic waves. These are waves of energy that travel through the Earth's layers and imparts low-frequency acoustic energy. And this energy is not only dissipated by the Earth, itself, but into its atmosphere as well.

An earthquake also consists of many low-frequency vibrations that occur simultaneously, which is why it sounds like a rumble of indeterminate pitch rather than a low hum. These vibrations are in the range of frequencies from about 1 Hz to 10 Hz. And there is enough of correlation or a link if you will, where scientists now believe vibrations induced by earthquakes can have great affect upon the ionosphere.

This correlation is in the form of scientific data that shows the vertical motion of the ground shaking and swelling from tsunami waves can produced vibrations, which pressed upward against the overlying layers in our atmosphere in a rippling effect like water.

Particularly in the ionosphere, these waves can be **amplified to thousands of times their original sizes**. These in turn create *electromagnetic* and weather disturbances to the point where radio and satellite communications are all but useless.

In the above example, the Earth acts as a giant *microphone*. This is why light flashes are sometimes seen (caused by static electric charge) as a result of an earthquake. Now we come to HAARP…and the reverse affect.

Like a quartz crystal, HAARP bombards the ionosphere with High Frequency that in turn heats the *auroral electrojet*, producing the snail's-pace ELF vibrations…say around **2.5 Hz**, which will then saturate and undulate the Earth beneath it like a giant *speaker* with this vibration at the exact same frequency. Let's keep that in mind.

But before we go any further with this, let's make a radical jump and go to the collapsing of a section of the Nimitz Freeway in Oakland, CA, during a 1989 earthquake. This was analyzed by geologist Susan E. Hough of the U.S. Geological Survey, who showed that the layer of mud beneath the highway resonated about **2.5 Hz**, and had a width covering a range from about 1 Hz to 4 Hz. This was enough to turn it into a quicksand-like substance that could flow like water.

Ms Hough further demonstrated that the earthquake, although it had a mixture of frequencies, **strongly resonated at 2.5 Hz**. And the mud below the highway responded robustly to those vibrations with its own natural **2.5 Hz resonant frequency**. But the overpass, itself, also had, by coincidence, **a resonant frequency of 2.5 Hz** as well. Needless to say, this seismic vibration at **2.5 Hz** was the direct cause for sections of the bridge to collapse…and killing a number of people.

Now let's fast-forward to March 11, 2011. It has been widely known and reported that HAARP broadcasted a signal that created a **2.5 Hz** ELF frequency. This was recorded from 0:00 hours to about 10:00 hours…or for 10 hours prior to the Japanese earthquake, which had a magnitude of 8.9 on the Richter Scale and ensuing tsunami. This earthquake and tsunami then took the lives of over 15,000 people!

If you were to have gone to HAARP's official website (which is now down), you would have seen for yourself that the **2.5 Hz** frequency wasn't only being broadcasted for 10 hours, **it was constantly being broadcasted for 2 days *prior* to the earthquake.** (Broadcasting began on March 8, 2011, just before midnight.) Although we don't have a copy of that, we do have a handy copy right here of the 10-hour broadcast on March 11, 2011. Take a look and judge for yourself.

### HAARP and the Japanese Earthquake of March 11, 2011

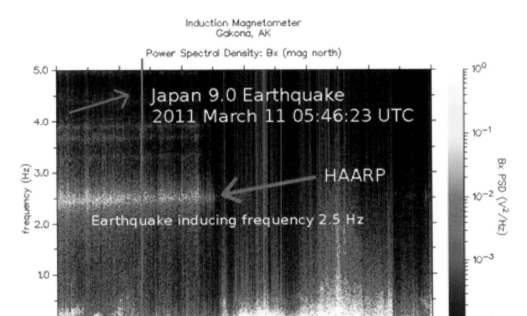

By the way, this magnetometer setup is a confirmation of their true intentions. If you're looking for a reflection, why isn't there a receiver set up somewhere looking for those High Frequency signals instead of a magnetometer looking at the induction waves of ELF frequencies from 0 to 5 Hz? In other words, they knew these ELF affects of 0 to 5 Hz were going to be generated and were actively searching for them *(and not the High Frequency signals)*, which had to have been deliberately predetermined and precisely calculated out previously! Need I say more?

But that's not the end of this story! The European Parliament issued a document that has defined HAARP ***as a weapon*** and has further placed NATO, the US Air Force and Navy on notice, demanding an explanation about their involvement in such disasters. Now this is not some sort of conspiratorial ranting or raving from a lunatic, but an official governmental inquiry with added protest, voicing legitimate concerns over this terribly new technology. It partially states:

*"The European Parliament...considers HAARP (High Frequency Active Auroral Research Project) by virtue of its farreaching impact on the environment to be a global concern and calls for its legal, ecological and ethical implications to be examined by an international independent body before any further research and testing; regrets the repeated refusal of the United States Administration to send anyone in person to give evidence to the public hearing or any subsequent meeting held by its competent committee into the environmental and public risks connected with the High Frequency Active Auroral Research Project (HAARP) programme currently being funded in Alaska;"*

The document can be read in its entirety at:

*http://www.europarl.europa.eu/sides/getDoc.do?pubRef=-//EP//TEXT+REPORT+A4-1999-0005+0+DOC+XML+V0//EN*

I wonder if the United States government would like me to represent them and explain HAARP and their position to the European Parliament?

## FACT: Invisibility & Cloaking Now Possible

The Lawrence Berkeley National Laboratory at the University of California recently created an ultrathin invisibility cloak – a thin film of magnesium fluoride **topped by a small gold antennae** – that can flexibly wrap light waves around any shape...to create illusions or to match different backgrounds. It does so by controlling the scattering of reflected light, and therefore, controls what the viewer sees. The cloak's creators say it can be draped over any object, obscuring it or making it appear as something else.

Cloaking with specifically engineered, artificial materials to bend light waves—*or metamaterials*—also holds great promise for electromagnetic field cloaking. "For more practical cloaking that can make large objects disappear for the human eye and work for all visible colors (frequencies), we think using standard optics (lenses and mirrors) has a lot of potential," said Joseph Choi, a researcher with the University of Rochester's Institute of Optics.

# Chapter VII: Beyond Earth - Mars

## *The Fascination of the Red Planet*

Even before mankind could point a telescope into the night's sky, humans have been mesmerized by spell of the glowing orange-red orb once called the "Devil's Eye." The planet Mars has filled our minds, our movies, and our myths with the realm of possibility of alien life…as well as the chilling vision of our own destruction. And it is just as popular a subject of conjecture today as much as UFOs and alien life are. Yet, it has always been so entangled with them.

There is no doubt that Mars holds some special deeper meaning to us humans…a meaning that is so engrained, yet obscured that we still cannot fathom it. Why do you think so many people believed that the Orson Welles' broadcast of October 30, 1938; a re-telling of "*War of the Worlds*," was the real deal? Thousands did…as well as the incredible march of the seemingly unstoppable invaders from Mars. And similar broadcasts, as late as the 1990s, have created the same type of panic and hysteria regardless of the huge number of warnings given throughout the theatrics of these shows of the reality of the content. Deep down, I do think that we all want to believe that there is life on Mars, with some ready to accept the slightest prospect as proof positive. This, however, is wrong.

This is not to say that we should discard any such notions or genuine observations of a real connection between the Red Planet, UFOs, and alien life. As an example, NASA's latest rover in 2014 photographed by what appears to be "a small flower" growing in Martian soil. Impossible? Not really. NASA scientists in 2011 (from photographs taken by the *Mars Reconnaissance Orbiter*) discovered new evidence that briny water flows on Mars during the warmest months. "This is the best evidence we have to date of liquid water occurring today on Mars," said Philip Christensen, a geophysicist at Arizona State University in Tempe. It raises the chances of life…including, I presume, the growing of flowers…which has our space experts and community, for some reason, stumped.

**Question:** *Are all of our NASA scientists in some sort of agreement at all as to what this seemingly plant life could or could not be?*

**Answer:** As always, each scientist has their own theory as to what exactly has been photographed. However, there are those who believe it is not a flower at all, simply because...*it resides on the planet Mars*...and therefore, is not a real flower. Talk about lack of rationale and deduction, real science always begins with the openness of mind and extends through curiosity, observation, and examination. How can one draw such a conclusion without examining the object, holding it in their very hand, itself? These individuals are either a major embarrassment to our space program or shills for a government that doesn't wish to share information with us...you know...**We, the People**.

## *The Lost Martian Space Probes*

Of all the places investigated by mankind, the exploration of the planet Mars and its two moons have had the greatest amount of missions conducted. However, according to the International Mars Mission Log, they also had the greatest amount of failures. With an overall success rate of 42 percent (51 attempts as of this writing), the space between Earth and Mars is considered to be the solar system's Bermuda Triangle presided over by the "*Galactic Ghoul*." This mythical entity coined by a few NASA scientists due to the concurrence of all these failures has been having a real field day.

However, if there is such an entity, he's been given a helping hand. On one previous NASA mission to Mars, a section of the launch vehicle that had been inspected and certified as clean, showed a "red light" on one of the mission boards. Upon re-inspection, it had been found that *an assortment of trash and dirt* had been dumped into the section, which surely would have resulted in a catastrophic malfunction. No question it was deliberate and of earthly origin. But why? Who wanted to see the mission fail,...a disgruntled NASA employee perhaps...or someone who was just following orders of government sponsored sabotage? Needless to say, the culprit still remains at-large.

In the case of the Russian spacecraft, *Phobos II*, however, this remarkable incident stands at the opposite end of the spectrum—as truly bizarre. Bizarre being some *unknown large object* of an elliptical design (saucer shaped), which after approaching the probe from the rear, apparently destroyed it in flight. The shadow of this object was seen against the Martian moon Phobos, which the probe was orbiting as the last of its camera's images were taken and returned seconds before the probe's unexpected destruction. As you can guess, no one has been able to satisfactorily explain this one away.

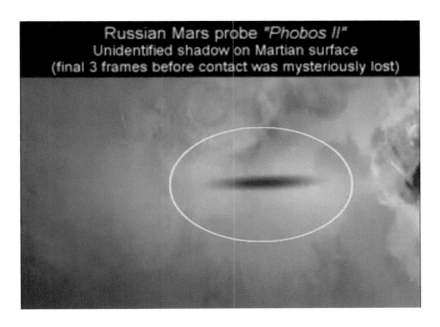

In between each success, a mixture of failures span a wide range of definitions. Launch failure, failure to deploy, communication failure, losing all contact with probe, probe crashing into Mars, and wrong trajectory are just a few of the problems encountered during these lost missions. Granted, some of them are genuine due to equipment failures and calculation errors as well as accidental things such as hitting unmapped space debris. But 60 percent is still a rather large number and hard to believe as purely coincidental.

**NOTE:** *In 2011, Russia's Phobos-Grunt spacecraft was launched. Its mission was (after an eleven month journey to Mars) to land on the Martian moon, Phobos, and collect soil samples and perform other experiments. However, it never left the Earth's orbit. Although the launch was successful, the probe did not perform its schedule burn and then all communication was lost. The probe finally broke apart and burned up in the Earth's atmosphere. Although there is no way of telling what happened to the probe that all of a sudden malfunctioned, the Russians have voiced the opinion that the United States' X-37 hypersonic spacecraft intercepted the probe and jammed its electronics after* *being guided to it by America's Space Fence (radar satellite tracking system). Interesting.*

**Question:** *If true, why would the United States interfere with another country's peaceful mission to Mars? Then again, why would any NASA technician fill a vital section of an American spacecraft bound for Mars with an assortment of trash and dirt so it too could malfunction? Makes you wonder, doesn't it?*

### The Case of the Mars Bio-Station

We grew up in the 1950s and early 1960s. Broadcast television, at that time, was the big thing, and color TV only affordable by the well-to-do. It was the era of family togetherness, when families sat in front of "the tube" and watched shows like "Andy Griffin" together. (Like most families, we could only afford only one black-and-white television set.)

During March of 1965, a most curious thing occurred after *Mariner 4* sent its very first pictures of the surface of the planet Mars back to Earth. Around eight O'clock in the evening, a news flash came over the airwaves at that point of time, claiming that a photograph of something "*as big as a battleship*" was taken on the surface of the planet Mars! These were the actual quotes I remembered as being given On-Air.

My brothers, my sister and I begged my parents to stay up and watch the news. My parents were just as curious as we, and so they relented. (We always went to bed at eleven O'clock back then.) So we waited until the news finally came on. After it did, the wait continued,…and continued,…and continued, until finally the broadcast was over. Nothing again was ever said throughout the entire broadcast (or the following nightly broadcasts the rest of that week) about the mission to Mars or the photograph of something "*as big as a battleship.*" Yet, we all heard that news tease of that evening…and were all just as puzzled. The mission was never mentioned in these broadcasts or the photograph ever shown. And it made me personally wonder for the longest time,…why? What did they think they saw? Why didn't the news anchorman mention anything about it during the broadcast? Was it pulled from the broadcast deliberately because it was a mistake?

Back then we believed in our government…believed what they told us. That is hardly the case now—some sixty *wiser* years later.

Fast forward to May 28, 2011. An "armchair astronaut" by the name of David Martines uploads a video to *Youtube* and creates an immediate sensation. Using Google Earth, Martines reveals a tubular structure on the planet Mars with red, white, and blue stripes on it.

From a distance, it looks nothing more than a white splotch on an otherwise unblemished red landscape. However, when zoomed-in upon, the details become more than evident. The object is estimated to be over 700 feet long by 150 feet wide using the measuring tool of Google Earth. Several news reporters even comment that it was "*as big as a battleship.*" Martines went as far to list the co-ordinates 49'19.73"North 29 33'06.53"West so others could go see the anomaly for themselves.

And just as immediately, the skeptics attacked the Martines video using their regurgitated verbiage from their standard playbook, despite the fact that these pictures are courtesy of NASA. Their chief debunker, a NASA planetary geologist at the Lunar and Planetary Lab at the University of Arizona and the director of the Planetary Imaging Research Laboratory, stated that this abnormality was the result of a "*linear streak artifact produced by a cosmic ray.*" **(Photograph following.)**

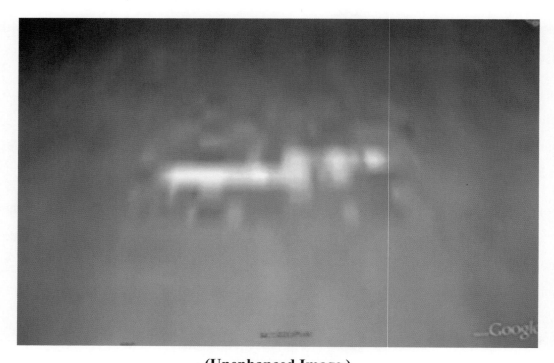

**(Unenhanced Image.)**

**(This is the NASA Image of the Bio Station. Does this really look like a scratch?)**

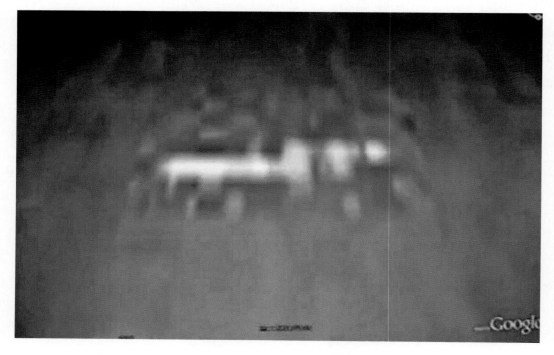

**(Enhanced Computer Image with Contrast increased slightly and Brightness decreased slightly.) As you all can see, the "Bio Station" appears quite real...*along with the complex of shadowy smaller buildings behind it!***

*Is that so?* I wonder if this NASA planetary geologist ever heard of a German physicist by the name of Heinrich Rudolf Hertz, a pioneer in radio communications. A unit of frequency (used in radio communications) was named after him. Called the hertz, this unit represents how fast an AC (Alternating Current) signal oscillates…or *alternates.* (*A sine wave known by another name.*) One hertz (Hz) equals one complete cycle per second. For example, 1000 hertz or 1000 cycles per second is equal to $10^3$ or 10 x 10 x 10. Alternating voltage in a wall socket (120 VAC) (Voltage Alternating Current) oscillates at 60 Hz or 60 times per second. I hope that is simple enough for even a NASA scientist to understand.

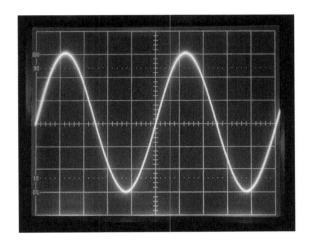

**A sine wave as presented on an oscilloscope.**

Since I was trained as an electronic engineer, it occurred to me that cosmic ray frequencies (**$10^{19}$ Hz [Hertz]** *and above*) are far faster than x-ray frequencies (**$10^{18 \& 17}$ Hz**), or otherwise put, 10 to 100 times faster respectively than x-rays. So what about the visible light spectrum? Well, cosmic ray frequencies are quite far above the visible light spectrum of frequencies, with visible light occurring at around $10^{15}$ Hz. This translates to cosmic rays being ***at least 10,000 times faster than visible light***. That begs some questioning, and some peculiar problems in both physics and mathematics if we were to believe this particular representative of NASA.

How is it possible then that equipment designed to receive lower frequencies of light can intercept a cosmic ray? And where would this cosmic ray come from? First there was the enormity of the planet Mars blocking any direct path of a cosmic ray approaching head-on to the camera lens as it apparently sat *below* the planet's horizon when the photograph was snapped. Only by a steep angle was accessibility to the lens at all possible. And by such a steep angle, shouldn't this type of effect be spread across a considerably larger portion of the entire picture area...*or would it have been circumvented by the lens cover?*

And although a linear appearance (rows of straight lines) would be consistent with a streak, the image itself *is not*. This is because many of the object details are at right angles to each other, which means *that two cosmic rays at right angles to each other had to impact the lens at the same point at the very same moment of time and at the very same frequency and amplitude*—if that happened at all. And what are exactly the odds of that? From a statistically point of view, it would be astronomical!

Oh, by the way, the two beam effect can actually be seen on a piece of electronic test equipment that has been around and *in use for over eighty years*: the oscilloscope. The oscilloscope utilizes two electric beams: one vertical and one horizontal. The vertical beam represents amplitude while the horizontal beam represents frequency and time. Together (when a signal is inputted to the device by means of a probe), 2D electronic analog and digital signals can be seen and analyzed when troubleshooting things like a TV or radio. However, if one of the beams was not working, all you would see is either a single horizontal line (trace) on the screen or a single vertical one with no depth to dimension by either. This proves my point about two cosmic rays at right angles having to hit it simultaneously, because one would not see right angle details contrarily. Which now dismisses the NASA planetary geologist's cosmic ray theory as illogical.

It is also curious that no other picture taken of Mars has this type of...if you believe it or not...*artifact* effect. With so many cosmic rays around, how come there aren't scores of pictures similar to the Mars Bio-Station? One would think all the photographs would be filled with them.

**Question:** *How can equipment designed for lower frequencies pickup and respond to a wave frequency at least 10,000 times faster?*

**Answer:** It can't because of the limits of its range of design and built-in electronic filtering processes. This is true of all *well-designed and shielded* equipment. If this wasn't the case, why hasn't the NASA planetary geologist produced any such *controlled-examples of his own* of this type of phenomena, demonstrating that the Mars Bio-Station can be replicated *at will* as proof positive of his claim? It seems to me that the burden of proof is on him to do so since he is the professed expert attempting to debunk the image. Where is his proof then? I just gave you mine, **which is absolutely verifiable!**

Surely with his great expertise and the state-of-the-art equipment he has at his command, he can do this so very easily or has some past pictures of artifacts that look similar to the Bio-Station. And I would be most interested in seeing any *unedited examples* as probably everyone else would…and not the Youtube *dubious* substitutions claiming that it was supposedly generated from ice crystals and minerals.

But let us assume he can't. And if he can't, that means that the image of the Mars Bio-Station was created by other means…like from a proper functioning camera, taking an actual picture of an existing structure on Mars!

And to support my views even further, I cite the human eye as evidence. The human eye is a very diverse piece of visual equipment for producing imagery from visible light only. But you cannot see a cosmic ray with your eyes, can you? Do you know anyone who can see a cosmic ray with their own eyes? Of course not: the human eye does not possess the capability of seeing cosmic rays. So think of it in the terms of a camera. If it is not designed to receive higher frequencies, how can it produce them then. The only way a higher frequency beam can interact with such equipment is to damage or destroy it (like a high power laser)…and it would be permanent. In humans, we would call it a burn or cancer, but unlike a camera, we can heal ourselves. And as a postscript, the camera which took that picture did not suffer any damage…and functioned perfectly throughout the rest of the mission. Does NASA have any rebuttals to that?

However, there is another way to prove what I stated above is correct. How? Simple, do this little experiment of trying to take a picture of a lightning strike during a storm using a cell phone or any other type of camera. And what you'll come to realize is that NASA planetary geologist was being very mistaken.  And all you'll have to do is play with the shutter speed to alter exposure time: *the time it takes the shutter to open and close.*

With your camera set to a normal shutter speed (which is very short and very fast), you'll get a picture like the one below.

*Your camera set to a very short (fast) shutter speed will create something like this image.* ***NASA satellite cameras are set for a short shutter speed also...in order to get high resolution and definition in their photographs.*** *So, that little fact in itself defines the Bio Station as being valid!*

Now, set the camera for a long (slow) shutter speed and take a picture of a lightning strike again.  The picture below is what you'll get.

What the long shutter does, is introduce the blur of the ball of lightning as it travels to the ground...the same thing your own eyes will see.  ***And the results of this experiment stands in direct contradiction to NASA's explanation!***

This little demonstration of the use of the shutter speed CLEARLY confirms and proves beyond a shadow of a doubt the image of the Bio Station is a factual, a solid object built on the surface of the planet Mars!

If that particular NASA planetary geologist cannot state with 100 percent *absolute certainty* what really created that image, then how can he declare it doesn't exist? However, I have proven the Bio-Station is a reliably solid object while reducing the NASA planetary geologist's explanation to a lack of understanding of both how cameras work while ignoring scientific principles.

The image speaks for itself. It is a clear visual object that has all the proper shadowing and distinct colors, giving it a 3D appearance. And since it was taken with NASA equipment, there can be little doubt as of its authenticity. Only the true size is debatable. If one uses the arc of Mars rather than the measure tool on Google Earth, we get an object somewhat smaller (by about almost half). Yet, still that is rather large, don't you think?

And there is one more matter to be considered here, its location from the northern polar ice cap. If you were to build a space station on Mars, wouldn't you put it close enough to a polar cap to give you a continual source of fresh water?

**NOTE:** *Fresh water is thought to be frozen below a mixture layer of solid frozen carbon dioxide and water ice on Mars. Make sense? Yet, it's far enough away so as to not be effected or threatened by the ice caps, which both expand at times and contract as well.*

All this beckons to my mind of my youthful recollection of a news broadcast way back in 1965. Was this the object that they saw? Was this the photograph of something *"as big as a battleship"* they reported?

A curious footnote to this whole affair: within two days of posting his video, Mr. Martines removed it, but by then it had gone viral. So several days after that, he aired something that sounded like a half-hearted explanation and apology. The question is, why? Why would he do such a thing…unless someone, possibly from the government, suggested it would be in his best interest to do so. (*Remember how the people of Roswell, their newspaper and radio station workers said that military officers and other representatives of our government threatened them in order not to have the flying saucer crash revealed? Maybe someone suggested if the video of the Mars Bio-Station wasn't removed, there was a place in the desert for Mr. Martines' bones as well.*) We'll never know. But rather than make Mr. Martines a martyr or have him hurt in any way, I think it would be best served to leave him to his peace.

## The "Face" of Cydonia

Of all the enigmatic controversies that surround the planet Mars is the many photographs taken of the huge earthen…or perhaps *constructed* faces that have appeared in different regions of the planet. In particular is the "Face" of Cydonia. Ever since *Viking 1* pointed a camera at this feature of the Martian surface and transmitted the "Face" back to NASA in 1976, the possibility of Mars once truly being inhabited has brewed. And strangely, NASA has been trying to explain these abnormalities away from the very onset (each and every time) as a trick of light rather than conduct a true scientific investigation. I wonder where their spirit of adventure is?

Sending a rover over to Cydonia, as an example, would prove beyond a shadow of a doubt how the "Face" was created. This would be the logical approach, wouldn't it? Why hasn't NASA done so? Surely if one was truly curious about the physical formation of the planet, the possibly of life and the peculiarities of the planetary environment, this area would be of high value interest. Why hasn't it been done? Would someone from NASA like to explain this? Maybe it's because if a rover starts sending video back to Earth, us humans might just be getting a eyeful of "things" that would have only one explanation…*we are not alone!* There are other objects in Cydonia, which tend to point to this, like the strange sets of pyramids which accompanies the "Face." But we'll talk about that a little later on.

However in going back to the "Face," NASA had it re-photographed from a different lighting angle (by the *MGS* spacecraft – Mars Global Surveyor) as well as running the images through a high-pass filter *"enhancing"* software process. This filtering software, however, tended to blur features, making some totally indistinguishable. This is because all lower frequency elements (beneath the high-pass cutoff frequency) were discarded. They didn't bother to tell anyone this when they first released the retouched photographs to the public *to prove* their conclusion of it being just a "*trick of lighting*." Well, the "*trick of lighting*" was what JPL (Jet Propulsion Laboratory) did to the images to dupe the public!

Unfortunately, some savvy private citizens (*who professionally analyze images for a living*) caught onto JPL's little deception and not only secured the raw footage, but managed to get NASA to fess-up to what JPL had done. And most curiously when these private citizens ran the raw images through *their own software*, they came up with something contrary as well as puzzling—*a human face*. This was primarily derived from the presence of secondary facial features (eyebrows, pupils, nostrils, etc,) seen with clarity by the more powerful camera lens and to the change of the lighting angle.

To add further to the mystery, the "Face" of Cydonia *is almost identical in appearance and size* to one built by the Sumerians, with their "Face" located in what is now present day Iran. And this Iranian "Face" is the type (*a raised mound marking the site of an ancient city*) of Malik Shah. Several photographs were taken of it from an altitude of 1150 feet in November of 1937. The site has been dated back to several thousand years BC. Amazing! **(Two Photographs following)**

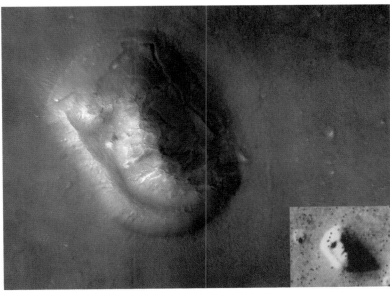

**(The "Face" of Cydonia)**

**(The "Face" of Malik Shah)**

**Question:** *How do you explain away two nearly identical structures built on two different worlds, Earth and Mars, millions of miles apart and before the dawn of prerecorded history?*

**Answer:** You can't. Here is proof positive that *Ancient Astronauts* did exist!

By the way, it is also interesting to note that JPL and MSSS (Mars Satellite Surveillance System) (the contractor for the MGS mission) initially refused to take the high resolution photographs of the "Face" on the stated grounds that it would be a waste of public funds as well as an insult to the scientists on the program. However, they were overruled by NASA headquarters.

**Rumor:** Some theorists suspect that NASA destroyed part of the Face of Mars with the Mars Observer, which blew up supposedly due to a faulty fuel line. This may give an accounting in photograph differences of the face between 1993 and 1998. The Mars Observer propulsion system used a high thrust, monomethyl hydrazine/nitrogen tetroxide bipropellant system for larger maneuvers and a lower thrust <u>hydrazine</u> <u>monopropellant</u> system for minor orbital corrections during the mission. These fuels can be highly explosive. However, unless evidence surfaces otherwise, such an act cannot be proven.

## Quoted from NASA Report:

*Because the telemetry transmitted from the Observer had been commanded off and subsequent efforts to locate or communicate with the spacecraft failed, the board was unable to find conclusive evidence pointing to a particular event that caused the loss of the Observer.*

*However, after conducting extensive analyses, the board reported that the most probable cause of the loss of communications with the spacecraft on Aug. 21, 1993, was a rupture of the fuel (monomethyl hydrazine (MMH)) pressurization side of the spacecraft's propulsion system, resulting in a pressurized leak of both helium gas and liquid MMH under the spacecraft's thermal blanket. The gas and liquid would most likely have leaked out from under the blanket in an unsymmetrical manner, resulting in a net spin rate. This high spin rate would cause the spacecraft to enter into the "contingency mode," which interrupted the stored command sequence and thus, did not turn the transmitter on.*

*Additionally, this high spin rate precluded proper orientation of the solar arrays, resulting in discharge of the batteries. However, the spin effect may be academic, because the released MMH would likely attack and damage critical electrical circuits within the spacecraft.*

The board's study concluded that the propulsion system failure most probably was caused by the inadvertent mixing and the reaction of nitrogen tetroxide (NTO) and MMH within titanium pressurization tubing, during the helium pressurization of the fuel tanks. This reaction caused the tubing to rupture, resulting in helium and MMH being released from the tubing, thus forcing the spacecraft into a catastrophic spin and also damaging critical electrical circuits.

## The Pyramids of Cydonia

Other equally puzzling structures in the Cydonia region of Mars is the nearby complex of pyramids. Although not true pyramids, they do have linear edges with relatively smooth triangular sides. This intriguing anomalous complex bears a striking resemblance to the pyramids of Giza in Egypt. And what makes these a greater oddity is that they are not only clumped together by the "Face," just like the Giza pyramids are to the Sphinx (also like the one on Mars), but the "City Pyramid" of Mars, a five-sided structure, resembles the five pointed Egyptian symbol for a star. And these pyramid-type structures not only show no evidence of having impacted or otherwise been deposited there, but are also found nowhere else on the planet. Another oddity is that both "City Pyramid" of Mars and the "Great Pyramid of Giza" are aligned to their respected planet's North/South axes. How does one make sense of that?

Something else to give thought to, are that triangular structures are not natural. I cannot name (or can anyone else for that matter) any natural triangular formations ever existing on the Earth. Granted the environments of Earth and Mars are different, but if winds and dust storms had carved one side of them, they surely would have destroyed any other side in the process by the shifting direction of wind and dust.

### The Other Faces of Mars

There is more than just the famous "Face" of Cydonia on Mars. The "Panda," the "Anteater," and the "Screaming Man" are craved within Cydonia's city limits and yet somehow are always overlooked. Another face, found in the Utopia region of Mars from photo #86A10 taken by the Viking 1 orbiter is an interesting find as well. And although the Galle Crater (a.k.a. "Happy Face" Crater) as seen by Viking in 1976 and during Mars Global Surveyor's 1999 revisit (Release #MOC2-89) may be of impact formation, I think the lavaly Kermit the Frog, frolicking on the Martian surface as seen by Viking 1 back in 1976 could use some explanation. Beyond the faces, there are also numerous "hearts" patterned on the Martian surface as well, at least a dozen. Incredible features such as winding glass-like tube with some sort of transportation car embedded within it or the different build-type structures rising out of the ground all over Mars add to the mystery. And there is so much more! All are a part of the enigma of the planet Mars…requiring further and honest examination.

### Martian Photographic Mysteries Taken by Curiosity

One of the biggest questions that has to be asked is: "Has the NASA rover *Curiosity* picked up real evidence of extraterrestrial life on the Red Planet?" There are several unique photographs taken by the rover *Curiosity* that demand further study.

**Photo #1: The Walking Lady** – Although some say it is a creature, it looks like a carved statue or statuette of a woman walking to me. It reminds me of a curio bought from an antique dealer selling Greek or Roman artifacts on a tour abroad.

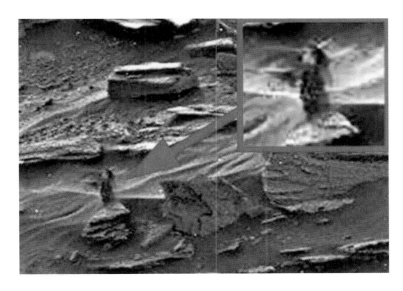

**The Walking Lady**

**Photo #2: The Fossilized Lizard** – The object has the appearance of a large walking lizard fossilized at the moment of its death. It reminds me of an old monster movie that substituted lizards for dinosaurs in fight scenes.

**The Fossilized Lizard**

**Photos #3: The Fossilized Lizard Rats** – These objects taken at two different locations have the appearance of lizard type rats that not only look exactly the same, but are in different poses. Would NASA care to explain this?

**The Fossilized Lizard Rats**

**Photos #4 and 5: UFOs** – This awesome discovery is not in only one photograph, but two NASA photos (Sol 369 and Sol 370), both of which show the same UFO in different positions.

**Airborne UFO over Mars**

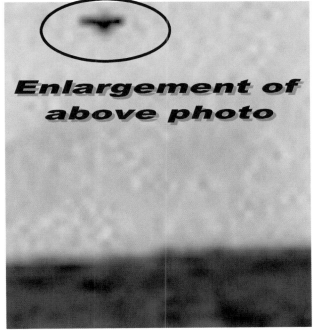

**Photo #6: Vanishing Martian City** – A hazy, distant object on the tip of the horizon mysteriously appeared and then disappeared in consecutive images of the Martian horizon, perplexing even NASA scientists. And it was large by any standard and kept the same features as it vanished. The official explanation was "a plume of dust kicked up by the sky crane that veered off and struck the ground some 2,000 feet (600 meters) away. Sure….

**The Vanishing Martian City**

**Photos #7: UFO** – On Aug. 7, two small white dots trek across the Martian sky in a time-lapse sequence were shot by one of Curiosity's hazard avoidance cameras, which NASA did confirm later. But after that, NASA seems to have "retouched" the original photos and then claimed these were actually shot on Aug 5th.

There were many other photographs taken by the rovers, which have not seen the light of day. However, there were some that barely did. Here are just a few of them:

**Structure dubbed "The Alien Base."**

 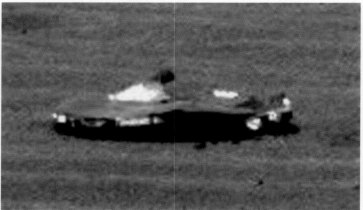

**Crashed Earth Vehicle Debris...or landed UFO on Mars?**

**Levitating Martian Rock...or an Illusion of Light?**

(I've seen a post that declares this as being a debunked visual illusion. My comment: "Ah...*no*—unless you landed on Mars and taken the time to walk completely around this anomaly *to verify such an opinion*.")

Apparently that person who posted the comment never heard of John Worrell Keely (an engineer of the 19th Century) and his scientific work on Etheric and Dynaspheric Forces. These forces can be readily produced in nature...or by someone with the knowledge and understanding...*of how to do it*.

*Proof Positive of this* can be found at a place called Coral Castle in Florida City, Florida. Between 1923 and 1951, a man by the name of Ed Leeskalnin single-handedly and without heavy machinery of any kind moved up to 1,000 ton limestone blocks without the aid of cranes or any other type of heavy equipment in creating the castle. Ed was born in Riga, Latvia in 1887 to a family of stonemasons, and immigrated to the U.S. sometime around 1913.

One hundred pounds, five-feet tall Ed, and a loner, he refused to let anyone even watch him work, carving or placing the massive stones...by himself—*by himself!* When asked how he manipulated such large blocks alone, Ed explained he had "*discovered the secrets of the pyramids.*"

These appear to be the same secrets of John Worrell Keely, who moved a 500-ton engine...*by himself*...as well as giving public demonstrations of other such astounding scientific feats. He too, kept this secret very well—and only now is his work being diligently researched and investigated to understand exactly what these two men had discovered and harnessed. It deals resonance and harmonic frequencies combining mechanical vibration, electric and magnetic fields, and acoustics with music. *And by the way, if you wanted more proof of **HAARP**, consider it as a crude, but an efficient utilization of these theories.*

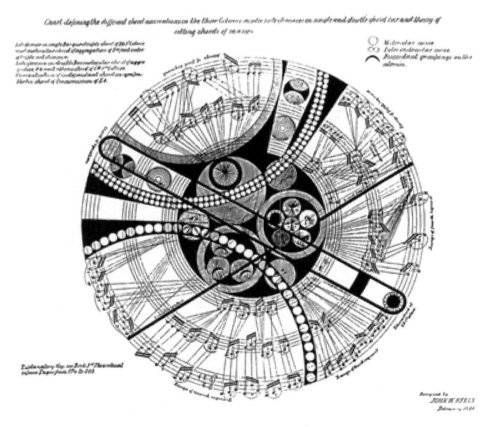

**One of John Keely's Diagram-Schematics (*Reminds you of Crop Circles, doesn't it?*)**

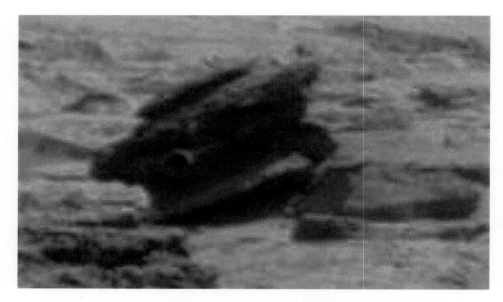

**Martian Truck Wheel Sitting Atop of other Vehicle Parts**

**Partially Buried Martian Hydraulic Valve**

**Another UFO over Mars**

**Very close Monolith UFO flying above Mars**

**(Maybe the Black Knight "space blanket" made its way from Earth to Mars?)**

It is curious that some may never be released from what I understand because of what they show, while others have had objects of interest airbrushed out. But this is, after all, hearsay unfortunately and cannot be taken as absolute truth. Others are dubious if not questionable when presented by conspiracy theorists, because some do show tampering, which hurts in qualifying fact. But there are those of extraordinary distinction like the examples above that do hold a key to the truth if one has the mindset and stoutness of heart to look for it. But the public will never truly find out until it steps foot upon the planet Mars itself.

## *Another Mystery of Mars*

**Reported on January 19, 2014**; a mysterious rock appeared in front of the *Opportunity* rover after being shut down for 12 days. With scientists at the Jet Propulsion Laboratory (JPL) in California monitoring *Opportunity's* images, astronomers were aghast that a large new "white" rock had "appeared" without any explanation on an outcropping "red" surface (and having a likeness of the Bimini Road), which had been totally empty just days earlier. **(Photograph following)**

**(Appearing Rock on Martian Surface)**

NASA issued a Mars status report entitled "encountering a surprise", and lead Mars Exploration rover scientist Steve Squyres said, "We saw this rock just sitting here. It looks white around the edge in the middle and there's a low spot in the center that's dark red – it looks like a jelly doughnut. "And it appeared, just plain appeared at that spot – *and we haven't ever driven over that spot.*"

Squyres said his team had two theories on how the rock got there – that there's "a smoking hole in the ground somewhere nearby" and it was caused by a meteor, or that it was "somehow flicked out of the ground by a wheel" as the rover went by. However, with no evidence of any "smoking hole" by a meteor that would have pelted and sprayed the rover with deadly stony debris (or for that matter a trail of pebbles), and with the rover having a maximum speed of just 0.05 mph (that's 4.4 feet per minute), both theories are more than a little hard to swallow.

There is also another curiosity about all the little pebbles surrounding the area of the white rock. If you examine all the pebbles in both photographs surrounding the "white rock" very, very closely,...you will see with your owns eyes...*they're all in the exact same spot!* **Not so much as one has been disturbed!** How can NASA explain this?

There is another thing that is most puzzling here, which even NASA pointed out; it's a "white" rock. This rock is not consistent with the terrain. It has the appearance of a sedimentary rock (which have been found in Martian dry lakebeds), possibly a combination of dolomite and siltstone. In other words, it came from another place that was wet at one time! The question is…what other place?

A sedimentary rock is a type of rock that is formed by material deposited on the planet's surface and within bodies of water. Sedimentation is the collective name for processes that cause mineral and/or organic particles (detritus) to settle, accumulate, and bind together in a liquid.

Mars once had, it is believed, a breathable atmosphere as well as being quite wet. But with the cooling of its core, it lost its magnet field. This in turn allowed the solar winds of our sun to strip it of atmosphere and water as well. What this means is that the rock is quite ancient.

But there is more strangeness to this rock; it's not cover in the red iron oxide dust of Mars. Surely anything exposed to the sandstorms of Mars would have a generous coating of Martian red iron oxide dust, rather than looking like it came out of a showcase. That means the rock was freshly picked up…and cleaned off *before being dropped in front of the rover's camera!* Someone has a sense of humor.

**Reported on April 2, 2014**; another image transmitted by NASA's *Curiosity* rover clearly captured an artificial light on the Martian surface and sent UFO enthusiasts wild about the possibility of life on Mars. The image is visible at the raw images database from NASA's jet propulsion laboratory:

(*http://mars.jpl.nasa.gov/msl/multimedia/raw/?rawid=NRB_449790582EDR_F0310000 NCAM00262M_&s=589*)

The upward illumination can be seen in the upper left-hand portion of the photograph. *Curiosity* snapped the image shortly after arriving at the "The Kimberley" waypoint on April 2, 2014. Over the past few days, UFO-spotting blogs have picked up the image as a sign that someone else...is very much out there! **(Photograph following)**

**Part of Original Photo**

Now, we were going to let NASA get off the hook with just the little blurb above and call it a day. However, a few of the NASA scientists decided to regale us again in (*what I deem to be, in my own opinion of course*) their total ignorance…or arrogance. Seems to me that by their tone and usage of their words, they think we're just a bunch of gullible and stupid people…and that they are so much smarter than the rest of us.

So, to you, the normal reading audience, please excuse the style of writing I am about to present in conjunction with this mysterious light. You see, in talking to these NASA scientists, I concluded that the only way they would apparently understand me and why this light is real and probably an indication of technology, is to talk to them rationally with facts in physics.

**NASA Skeptics' Theory:** From an unnamed NASA scientist: "This was caused by a glint from a rock's surface reflecting the sun."

**Rebuttal:** There are no other rocks in this entire series that demonstrate this same effect, and everything is cast in shadow. Another point to this is that the illumination crosses into a distant part of Mars…from a mountain twenty or so miles beyond. So how does a mountain twenty miles away reflect light evenly in combination with a rock that is fairly close to the rover? The answer is; it simply can't!

**NASA Skeptics' Theory:** From the same an unnamed NASA scientist: "photographic artifact resulting from the charged coupled device (CCD) that the camera uses to capture images."

**NOTE:** A **charge-coupled device (CCD)** is a major piece of technology in digital imaging. It translates received light into electrical charge for conversion into a digital value. In a CCD image sensor, pixels are represented by p-doped MOS capacitors. These capacitors are biased above a threshold for inversion when image acquisition begins, thus allowing the conversion of incoming photons into electron charges at the semiconductor-oxide interface. The CCD is then used to read out these charges. Although CCDs are not the only technology to allow for light detection, CCD image sensors are widely used in professional, medical, and scientific applications where high-quality image data is required.

**Rebuttal:**

All other pictures in this series are crystal clear and show no defects. Camera appears to be operating flawlessly. The CCD (a semiconductor device as described above) would have developed a "hole" in what is called its substrate material (different silicon layers "doped" with special chemicals and etched to produce an electronic device or circuit), in which case, numerous pictures would have the same defect also within their images. Ergo, it is not a CCD problem.

**NASA Skeptics' Theory:** From the same an unnamed NASA scientist: "The rover science team is also looking at the possibility that the bright spots could be sunlight reaching the camera's CCD directly through a vent hole in the camera housing, which has happened previously *on other cameras* **(but not this one)** on *Curiosity* and other Mars rovers when the geometry of the incoming sunlight relative to the camera is precisely aligned."

**Rebuttal:**

Well then let's see a couple of these (raw) pictures that you claim have an artifact created by the vent hole by this camera as well as some of the others. I'd wager they don't even look remotely alike.

**NASA Skeptics' Theory:** From several anonymous NASA scientists: "Cosmic rays smashed the camera's detector."

**Rebuttal:**

This is really a good one. If the camera detector circuit wasn't working, you wouldn't be seeing any pictures at all. Everything would be blacked out! It is statements like these that make me wonder just how these scientists got their degrees.

### *A TRUE ANALYSIS OF THE PHOTOGRAPH OF THE MYTERIOUS LIGHT*

First of all, I downloaded the "raw" image of the photograph directly from the NASA site. Please remember, using any type of imaging program, you can accidently generate a false image, which I tried very hard not to do. I used one filter…and one filter only, I did a simple "resize." I also did the "resize" within acceptable limits and also using several other imaging programs for comparison. And by strangest coincidence,…they all illustrated the very same effect. Directly below is that image…and it has not been manipulated or altered in any other way. From this, I made my analysis. **(Photograph following.)**

**(Image "Resized Only")**

A little different now, isn't it? What we have here is not just a light, but a gas ignition. It is reminiscent for me of seeing the afterburner of a F4F Phantom Jet during a scrambled takeoff for a military bombing mission some 50 years ago. Compare the picture above to the photograph below. **(Photograph following.)**

**(Military Fighter in Afterburner)**

**NOTE:** *An **afterburner** (or a **reheat**) is an additional component installed on military supersonic aircraft. Its purpose is to provide increase thrust, usually for either supersonic flight (for interception or combat), or takeoff. Afterburning is achieved by injecting additional fuel into the jet pipe downstream of the jet engine's turbine.*

**NOTE:** *When comparing the photographs above, one has to remember there are differences between the atmospheres of Earth and Mars. These differences will give some distortions when trying to compare photographs, but overall they'll appear the same. The photograph of the jet fighter was filtered to grayscale with a very slight increase in contrast due to the time of day it was taken.*

The first thing we notice in the Martian photograph near the bottom, there is intense area of burning occurring. (It is burning white hot.) This is the source of the ignited gas and makes sense since the Martian atmosphere is mainly **Carbon Dioxide** (a very heavy gas ***used in fire suppression*** by denying oxygen to the flames, which they need to burn). The lighter gases are trying to escape upwards as they continue to ignite within a contained oxygen bubble plume. There is also a "halo" affect around this bubble. This is caused by two different things:

\* the difference in temperature between the ignited heated gas bubble, and the cold atmosphere of Mars,

\* the burnt exhaust gas contained around the outer portion of the *air bubble*.

The rover only took one picture of that area, however. If the rover had taken another picture a second later, we might have seen a plume of dark smoke rising, which would have confirmed this. But that didn't happen…or at least NASA didn't post it.

Combustion needs three things: fuel, an oxidizer, and an ignition source. This is basic chemistry. So what causes a gas eruption and ignition on a world whose atmosphere is mainly Carbon Dioxide and Methane (Methane needs an oxidizer to burn by the way) on a planet that is also considered volcanically dead? Besides that, the ignition is extremely narrow…and seemingly focused and controlled because of such straightness. It is acting like some sort of relief valve or mechanism spewing possibly an over-accumulation of flammable gases into the atmosphere…or possibly some deadly Martian bacterial agent and igniting them to safely dispose of them…and that my friends smacks of technology.

**NOTE:** *How we launch nuclear missiles from submerged submarines is the perfect example of what happened on Mars. For a nuclear submarine underwater -- A column (or bubble) of air is released with the ejected missile from a submarine. The missile motors then ignite in the air bubble, and fire. The air bubble ensures that the sea water does not rush in and extinguish the burning of the rocket motors while it rises out of the sea, completing its underwater launch.*

Are there any further comments, NASA? If not, then please tell me why you won't investigate this. I was somehow under the impression you've been dropping probes on Mars (**with taxpayer money**) to study the planet and learn its mysteries and secrets. This just might be a good candidate to start with.

# Chapter VIII: Beyond Earth - The Sun vs Venus

## *Things Far Beyond Technology & Imagination*

There seems to be some confusion by some critics when a genuine UFO is spotted submerged within the sun or traveling through its corona. These same critics will then swear that it's some sort of phantom or "*ghost image*" made by the reflection of Venus. (*Venus has long been used as an excuse for many other UFO sightings including the death of Captain Mantell when his P-51D fighter fell from the sky chasing one. Refer to Chapter III of this book.*)

There are three reasons why that can't be. The first is the most obvious. Since the temperature at the surface of the Sun is about 10,000 degrees Fahrenheit or 5,600 degrees Celsius, if you will, any reflective material they we know of could not possibly exist in that kind of heat. For example, *tungsten steel melts at 6,192 degrees Fahrenheit*. (Also remember the internal temperature rises from the surface of the Sun towards where it reaches about 27,000,000 degrees Fahrenheit or 15,000,000 degrees Celsius at its center.)

The Sun is spewing three things: *glowing and extremely heated particles*, a huge magnetic field, and a deep gravitational well. So how could Venus reflect off of those?

Things like "*Sun Dogs*" require ice crystals in an planet's atmosphere to reflect light. You've got to ask yourself a question here, what happens to a ball of ice called "*a Comet*" when it approaches the Sun...*from millions of miles away?*

**Answer:** It heats up and leaves a streaming trail of melting ice crystals in its wake. So if there is no reflective material, what is reflecting the light then? The answer to that is simple, a light other than the Sun has nothing to reflect off of. **Strike One!**

Below is a photograph taken of a "Sun Dog" on our planet called Earth caused by ice crystals in our atmosphere. Oh, and by the way, ***there are no temperature inversions on the Sun either***—just in case some critic wants to use *a mirage* as an excuse.

**A Sun Dog - (The Sun is in the center of the oval halo.)**

**The second reason:** Reflected light will always be overwhelmed by an intense light source such as the Sun. The most simplest way to prove this is to shine an ordinary flashlight with a color filter mounted on its lens at some sort of intense light source like a spotlight as the one portrayed below. Granted that this is not the sun, but do you really think you'll see any inkling of your colored beam reflecting off of its illumination? *"Ah,...no!"* **Strike Two!**

This was actual proven in World War 2. Ever hear of the Canal Defence Light Tank? It was a powerful searchlight mounted on a tank, with a shutter allowing it to flicker six times a second. The 13-million candlepower searchlight was intended to illuminate the battlefield while dazzling and blinding the enemy. They also hide the vehicles themselves—which could only be seen *up-close* in the day, when the search lights did not illuminate the corridor of light it created. Please understand, other than something projecting light, everything you see around you including colors is reflected light. A material that is seen in reflected light absorbs a single light band of one particular frequency while reflecting the rest. That's the scientific principle behind how a spectrometer works. That's how NASA can tell from great distances what material makes up of a planet beyond our solar system.

*The angle of the beam dispersion from these search light mounted tanks were 19 degrees, which meant that if the CDL tanks were placed 30 yards apart in line abreast, the first intersection of light fell about 90 yards ahead and at 1000 yards. The beam was measured at 340 yards wide by 35 feet high.* **This formed triangles of darkness between and in front of the CDL's into which could be introduced normal fighting tanks, flame-throwing Churchill Crocodiles and infantry.** *However, these were only used by the British during the D-Day landings...and strangely at no other time during the war.*

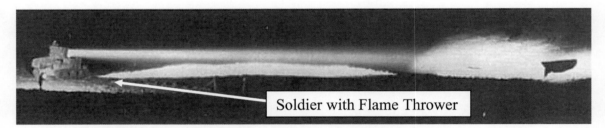

**A CDL Tank in action. (The tank cannot be seen at all by the target. Only in a side view with another light aimed near the tank's treads revels it along with the man standing by it and another with a flame thrower.)**

**The third reason:** the Sun's Gravity. An eclipse of the Sun revealed in 1919 and proved that light bends around our Sun—affirming Einstein's theory of general relativity. This was conducted by British astronomers headed by Arthur Stanley Eddington, Britain's leading astrophysicist who photographed the solar eclipse. Since they proved that gravity bends intense rays of light, what chance do you think a reflection could be seen as the Sun's gravity bends it in another direction? I'll tell you, zero to none! *Strike Three! The intellectual laziness of using "reflection of Venus" is OUT!*

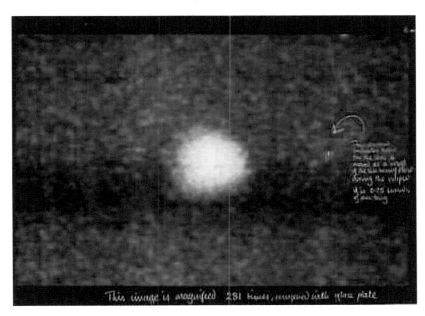

**An Actual Photograph of the bending of light by the Gravity Well of the Sun**

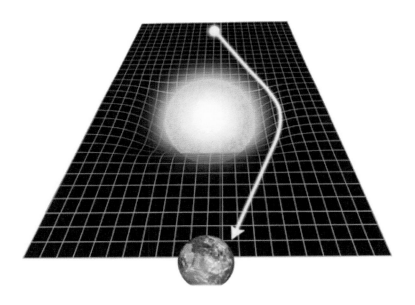

**The Actual Principle demonstrating the bending of light by the Sun's Gravity Well**
***ACCORDING TO EINSTEIN!***

As you can clearly see, any reflected image pointed at the Sun would be bent away exactly like the star's rays from behind the Sun by its massive gravity well! This one fact alone destroys the critics' argument of a "*reflection of Venus*" being responsible for UFO sightings.

So there you have it, the images you are about to see **ARE** solid, verifiable UFOs interacting with the Sun. Enjoy.

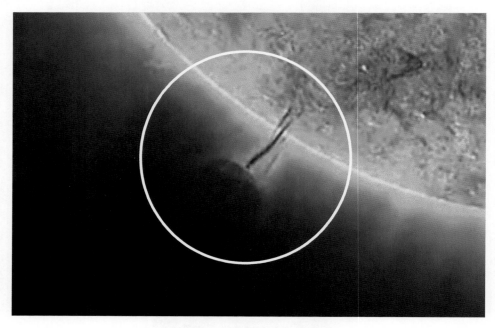

**NASA Photograph 2012:**
**Object appears to be drawing Material from the Sun.**

**NASA Photograph 2012:**
**Space Vehicle appears to be flying through Corona of the Sun.**

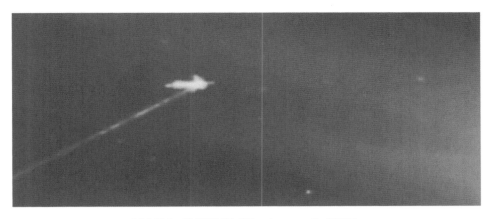

**NASA (SOHO) Photograph 2012:
Space Vehicle emits beam in Corona to the surface of the Sun.**

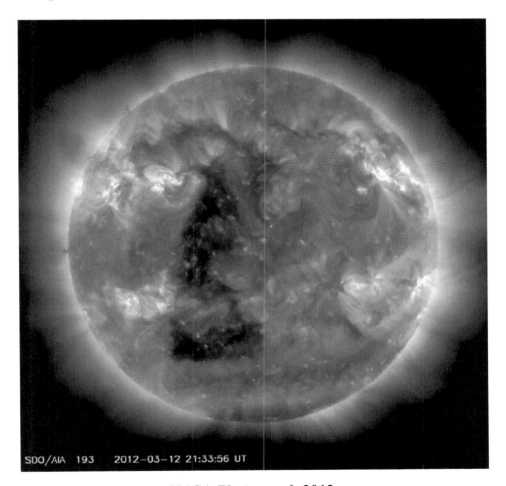

**NASA Photograph 2012:
Triangular Space Vehicle appears to be flying through depths of the Sun.
(Maybe it's Darth Vader's Star Destroyer.)**

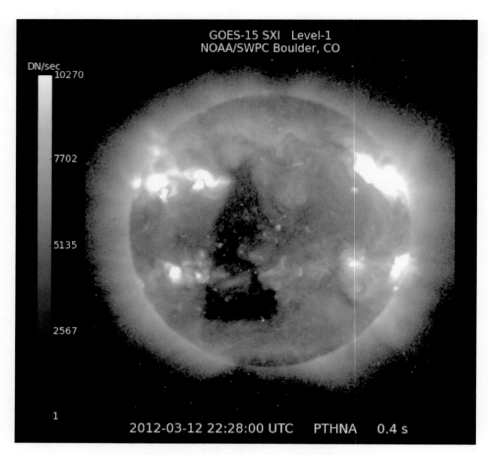

**Related NASA Photograph 2012 to above:**
**Triangular Space Vehicle appears to be flying through depths of the Sun.**

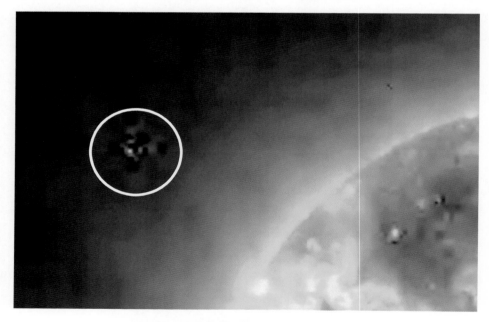

**NASA (SOHO) Photograph 2013:**
**Space Vehicle appears to be flying through Corona of the Sun.**

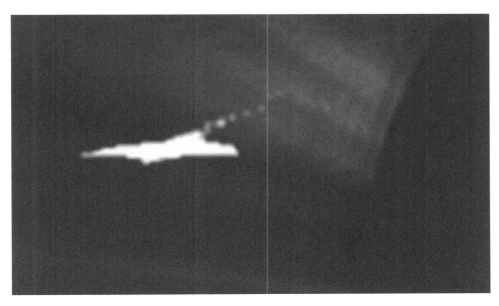

**NASA (SOHO) Photograph 2014:**
**Space Vehicle appears to be flying through Corona of the Sun.**

**NASA Photograph 2019:**

This Space Vehicle directly above appears to be flying next to eclipsed Sun. Due to the gravity well, this space vehicle could possibly be behind the sun, but it is by its technological shape still a space vehicle and not a distant star.

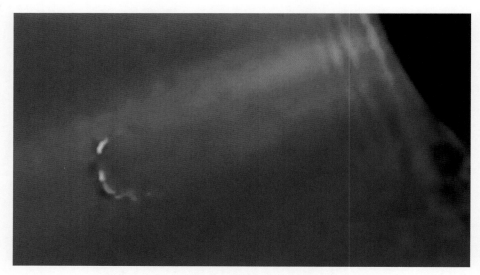

**NASA Video 2019:**
**Space Vehicle held position as Sun ejected a massive solar flare.**

If you ask any UFO critics using our explanations and pointing out that these are not any type of reflection,...*or refraction for that matter*, what do you think their "*rational*" justification for these space vehicles captured in these images will be? You'll probably only hear the sound of silence.

# Chapter IX: Misdirection, Hoaxes, Insufficient Data, Misidentifications and Misconception

## *Fact: The Twining Memo – Misdirection*

In September of 1947, a letter called "The Twining Memo" circulated throughout the Air Material Command regarding flying discs. **The United States Air Force at that time denied that this memo ever existed**, but this denial was proven unfounded when several copies of this memo were retrieved, which included one copy from the United States Air Force Adjutant General's file. (*Copies of the first two pages of the Twining Memo can seen in Chapter IV of this book concerning Roswell.*)

The subject of this memo was the authorization for the budgeting of a project to study UFOs written by Lieutenant General Nathan Twining. Apparently, critics of this document had obtained a draft of this memo to support their claims that this memo is a fake because the dating that appears on this draft was incorrectly displayed and there was no signature. However, proponents of this memo have rightfully pointed out since it is a draft, discrepancies would occur on versions of any unreleased draft memo, which are usually not signed at all. If this draft is a falsehood then how do these critics explain *Project Sign* and *Project Grudge*, which studied UFOs before the smoke screen they called *Project Blue Book* came into existence?

The "Twining Memo" was retrieve through the Freedom of Information Act. The preliminary study of UFOs was directed by Brigadier General George Schulgen and states that this research was performed by Army Air Force Intelligence and the Army Engineering Division. They suggested the conclusions listed below (in part) from this "Twining Memo" as follows:

*A. The phenomenon reported is something real and not visionary or fictitious.*

*B. There are objects probably approximating the shape of a disc, of such appreciable size as to appear to be large as man-made aircraft.*

C. There is a possibility that some of these incidents may be caused by natural phenomenon, such as meteors.

D. The reported operating characteristics such as extreme rate of climb, maneuverability (particularly in a roll), and action which must be considered evasive when sighted or contact by friendly aircraft and radar, lend belief to the possibility that some of these objects are controlled either manually, automatically or remotely.

E. The apparent common description of the objects is as follows:

1) Metallic or light reflecting surface.

2) Absence of trail, except in a few instances when the object apparently was operating under high performance conditions.

3) Circular or elliptical in shape, flat on the bottom and domed on top.

4) Several reports of well kept formation flights varying from three to nine objects.

5) Normally no associated sound, except in three instances a substantial rumbling roar was noted.

6) Level flight speeds normally above 300 knots are estimated.

## Fact: Gorman's Battle – Insufficient Data

October 1, 1948 – Fargo, North Dakota – 2nd Lieutenant George Gorman, a World War II Veteran Pilot, was in his P-51 Mustang flying night practice maneuvers at 21:00 hours. When a piper cub civilian aircraft passed 500 feet below him, he noticed an object to the west, but saw no wings or fuselage on it. The object appeared to have a blinking light. Lieutenant Gorman contacted Hector's airport tower in Fargo and asked about other aircraft in the area besides his own and the piper cub. They replied that their aircraft were the only two in the area. Lieutenant George Gorman reported the blinking light and the tower contacted the pilot of the piper cub and asked if he could see the blinking light. That pilot replied that both he and his passengers could see this light in the west.

Lieutenant Gorman stated that he was going in pursuit of the unknown. Accelerating to maximum speed of 350 to 400 MPH Lieutenant Gorman tried to catch the object, but it was out distancing him. He made a series of turns, circling the object trying to cut it off and close the gap between them. He did manage to gain on it, but it was now 500 feet above him. He could make out that the unknown was a circular object about 8 inches in diameter. The object then turned and headed in his direction. It then shot up to 14,000 feet with Lieutenant Gorman in pursuit. However, upon reaching 14,000 feet Lieutenant Gorman plane stalled allowing the unknown to reach 16,000 feet. Lieutenant Gorman made several more attempts to try and catch the object, but failed. In the course of the pursuit Lieutenant Gorman lost consciousness once, which allowed the object to escape his sight. At 21: 27 hours, he broke off the pursuit having lost visual on the object. Upon his return to base he stated that the object appeared to be under intelligent control.

The Army Air Force sent investigative officers from Project Sign to research this incident. They questioned Lieutenant Gorman at length for several hours. They also checked his aircraft for any radiation readings. They found that his aircraft did have a higher concentration of radiation on it compared to the aircraft that were on the ground. However, after checking further, they discovered that aircraft that are at higher altitudes will have a sizeable amount of radiation over the normal range. This was due to the thinning of the earth's atmosphere shielding the aircraft less from radiation at a higher altitudes. They also found that the AIR Weather Service had released a weather balloon on October 1st around 20:50, just ten minutes before Lieutenant Gorman's sighting. This balloon would have been near the airport when the object was sighted and prevailing winds would have given the balloon the upward and downward momentum. Thus this case was closed as a mistaken identity of a weather balloon.

### *Fact: George Adamski - Hoax*

In 1950, an individual by the name of George Adamski claimed to have contact with blonde haired aliens. He said that they took him to the moon where he took photographs of their space craft. The problem here is that how did he take these pictures of the exterior of their craft *from the interior of this craft*? Did these aliens use mirrors on the moon perhaps for creating the ambience for the photo shoot? So why didn't he take any pictures of these aliens and the interior of the craft as well?

**Question:** *Why did these aliens have to take him to the moon to take pictures of the outside of their craft, which strangely resembled the underside of a 4 bulb desk lamp? Maybe he needed a little light on the subject.*

**Answer:** As you can see, there are a few inconsistencies here. And it seems to me that such aberrations are always played up as a preemptive dismissal of any real eyewitness testimony being credible.

### *Fact: The Condon Committee – Misdirection*

In 1947, the United States Army Air Force started its investigation on the UFO phenomena with Project Sign. The population became greatly concern with the topic of UFOs and the government was afraid that this would cause a panic, changed the investigation to project Grudge trying to eliminate most of the reports as normal occurrences so as to tell the population that they had nothing to fear. This project was eventually molded into the famous Project Blue Book, which was openly criticized years later by its chief consultant and investigator Doctor J. Allen Hynek. On April 5, 1966 at a congressional hearing, Doctor J. Allen Hynek called for an independent civilian panel of scientist to examine whether UFOs were a major threat.

Walter Roberts director of the National Center for Atmospheric Research and Astronomer Donald Menzel were chosen by the Air Force to oversee who would get the grant for this study. They looked at several universities and chose Physicist Doctor Edward Condon from the University of Colorado to head the project with a grant of $313,000. Doctor Edward Condon felt that this was a small amount for the research on this project. The committee consisted of Doctor Edward Condon, Assistant Dean Robert Low, Astronomer Franklin Roach, Astronomer William Hart, two psychologists, one chemist, one electrical engineer and a graduate student. Two civilian investigation organizations NICAP (National Investigation Committee for Aerial Phenomena) and APRO (Aerial Phenomena Research Organization) opened their files for the Condon Committee, as they were known by, to review.

May of 1968, Doctor Edward Condon openly condemned the study as silly nonsense in "Look" a national magazine. This was done even before the research started. (So much for Doctor Edward Condon's unbiased objectivity.)

The Condon Report final conclusion after a two year study was that the existence of UFOs cannot be justified and recommended that no further study should be made on this subject. Doctor J. Allen Hynek stated that the Condon Report was slanted and that at least a quarter of these cases investigated were given inadequate explanations by the committee. Astrophysicist Peter Sturrock agreed with Doctor Hynek findings, stating in that all critical reviews of the report came from a subjective scientist, who had investigated and researched the UFO phenomena previously, holding a biased opinion.

In reviewing the outline of the Condon Report, it appears that this project was flawed from its beginning in that the head of this committee did not have an open mind with regards to the possibility of the existence of UFOs. There was also the problem with personnel chosen to work on the project in that their field of experience had little too no relationship to the subject matter being researched. Most astronomers did not believe in UFOs, and therefore, did not take any of this research seriously.

Furthermore, the two psychologists, one chemist and a graduate student had no experience or even knowledge to investigating Unidentified Flying Objects. We believe the committee should have been comprised of astrophysicists, astronauts, pilots, nuclear physicists, aerospace engineers and criminal investigators. But since the United States Air Force was paying for the research, this may well have been another reason as to why the outcome was in heavily favor of the government position that UFOs do not exist. Perhaps they should have renamed this study to the Con-Dumb Report for people who believe the government.

## *Fact: Billy Meyers - Hoax*

Billy Meyers in 1977 claimed to have direct contact with aliens from the Pleaidies Star System. He produced photographs of what he called Beam Ships. He also photographs occupants of these craft. He stated that they had made several visits to him over a period of months. Upon investigation it was noticed that one of these so called occupants resembled a well known Swiss fashion model and the Beam ship was discovered in his barn.

### Fact: Gulf Breeze – Hoax?

Gulf Breeze Florida is situated on the eastern seaboard of that state. There is a naval base, which has a restricted area located within the facility, and is across the bay from the town. In the 1980s, this town became flooded with reports of Unidentified Flying Objects. Many of the witnesses took pictures and films of these UFOs. There was one short film that was put on the internet of a UFO going behind some trees. UFO researchers went to Gulf Breeze to investigate these sightings and procured these photographs along with some of the films.

This evidence was submitted to several photographic experts. One expert said that the photographs were indeed of genuine UFOs. Another photographic expert stated that they were faked. One set of photo's appears as if the cameraman shot the pictures at a window, which held a reflection of a lamp. There was also the testimony of a woman and her son age 10 that a UFO was seen from their backyard. The boy said that he and a friend were playing when they looked up and saw the craft at a distance through the trees. They ran into the house and told his mother who wanted to see if they were telling the truth. She sat them down at either end of the table and had them draw what they saw. To her surprise both the pictures matched. This story appeared on the television special "UFO Cover-UP Live".

This Gulf Breeze incident does not mean that the UFOs seen there were real or that the witnesses were lying. A television show that addresses the paranormal called "Fact or Faked" decided on one of their episodes to investigate this incident.

Pulling one of the videos off of the internet of a UFO that was going back and forth behind the trees, they decided to try and replicate this incident. They then went to Gulf Breeze and found the exact location where the footage was filmed.

Going to a local arts and craft store, the TV crew picked up some inexpensive materials for their reproduction. Building an almost exact replica of the craft, they made several attempts of recreation the footage. Finally on their last attempt, they matched the video exactly that was seen on the internet by using the GOBO Test (Go Before Optics). This was done by placing a black screen behind the trees and projecting an image of a UFO onto the screen. Thus the person or persons viewing the image on the other side of the trees only sees that image moving back and forth making them believe that the object being viewed is actually there.

Afterwards, the TV crew ran the observer's vocal testimony against a vocal deception test and his testimony failed indicating his statement about the UFO was deceiving and may have known more about a hoax being perpetrated. It became apparent that some witnesses were duped in some of the reports. This, however, does not account for some reported sightings, which appear to be genuine. However, many of the still photos and other films can now be attributed to hoaxes or misidentifications.

## *Fact: Cattle Mutilations - Misidentification*

In regards to some UFO incidents from the mid west and mid southwest there have been reports of livestock mutilations. Sightings of strange lights in the night sky have been seen, usually a day to several days before the bodies of these animals are found. These mutilations have mainly been restricted to cattle, however some horses have also been found in the same condition. Herd animals (herbivores) are generally skittish at night and are always on the alert for predators. The History Channel "UFO Files" sent some of their investigators out to research these mutilations. They tried to approach a cattle herd at night, however; each time they got close to the herd the cattle moved away from them.

In the investigation, the mutilations always seemed to have a precise incision as if burned by a laser. Parts of the animal would always be missing. UFOs seemed to be the ranchers and ufologists prime suspect, however devil worshippers and psychotics (and possible several agencies of the United States government) have also thought to be the cause of these incidents.

**Speculation:** A television show called "Fact or Faked", whose premise is to investigate unknown phenomenon, decided to research these mutilations in one of their episodes. They went to a ranch in San Lucs, Colorado where a recent mutilation was discovered.

First they spoke with Mister Sanchez the owner, who stated that several nights before he had seen lights in the sky by the distant mountains. Two days after this he found one of his calves mutilated. There was no blood in the animal and its internal organs were missing. The very next day the calf's body had vanished.

The "Fact or Faked" research team contacted a local veterinarian named Amy Mason to help with the research. She had a cow that had died of natural causes brought to the ranch. She then proceeded to dissect the animal using several different methods. The first method was to try cauterization, which involved cutting the animal with a torch like device. However, the device took too long to make an incision and seared the wound stopping the blood loss. The second method tried was to use a laser; however this method for the mutilations was discarded as it needed an electrical source to run the laser.

The laser also had a similar effect on the dead animal as the torch did in that there was also no blood loss because the wound was immediately cauterized by the laser and it also took an extremely long time to make the incision.

The third method was to try a scalpel, but this too failed as the incision was too small and would have taken all night. Their conclusion was that the incisions were not man made. They then decided to examine if these incisions were from a natural cause. They covered a cowhide over a sturdy balloon. They then inflated the balloon and found that as the balloon inflated, much like a bloating internal organ of a dead animal; it caused a precise cut into the cow hide as seen in the mutilations. Prior to this experiment the researchers had placed a camera above a dead calf, and upon viewing the footage they saw several predators attacking the remains. It was concluded that the mutilations were from natural causes.

**NOTE:** *Herbivores have a natural tendency to hide any ailments so that predators do not see them as easy prey.*

Thus a rancher may not have noticed that any of his livestock were ill. Therefore it can be surmised that the blood loss and damage to the dead animals was do to predators drawn to the carcass by odor after the animal had expired.

## *1945 December, Florida: Flight 19 - Misconception,...yet a Mystery*

**Grumman TBM Avenger torpedo bomber: type of aircraft flown by Flight 19**

On December 5th, 1945; a group of five Grumman TBM Avenger torpedo bombers with the call sign of *Flight 19* went on a routine training mission, to vanish into legend. Over the many years, the public imagination has been captured by Flight 19 and the enigma of the Bermuda Triangle. As scarce answers and the US Navy's official stamp of "***top secret***" overshadows the incident, we done our best to shine the light of truth upon this mystery.

Some individuals truly believe that Flight 19 was abducted out of the Bermuda Triangle by UFOs and was even put into the movie "*Close Encounters of the Third Kind*" Being amateur researchers, we decided to tackle the Flight 19 case to explore all the possibilities and merits to this claim. We, ourselves, don't believe that any alien encounter occurred. However, what we found was a greater mystery.

First of all, one of the aircraft of Flight 19 *might have really been found.* This is in regards to Aircraft Designation: FT-36, with Tail Number 46094. The aircraft was flown by Pilot - Captain E. J. Powers USMC, Ball-Gunner - Staff Sgt Howell O. Thompson USMCR, and Navigator/Radioman/Gunner - SGT George R. Paonessa USMC (*the center of controversy in this story*). Below are the complete listing of all aircraft and crews of Flight 19.

| Aircraft Number | Pilot | Crew | Bureau Number |
|---|---|---|---|
| FT-28 | LT Charles C. Taylor, USNR | AOM3c George Devlin, USNR<br>ARM3c Walter R. Parpart, USNR | 23307 |
| FT-36 | CAPT E. J. Powers, USMC | SSgt Howell O. Thompson, USMCR<br>SGT George R. Paonessa, USMC | 46094 |
| FT-3 | ENS Joseph T. Bossi, USNR | S1c Herman A. Thelander, USNR<br>S1c Burt E. Baluk, JR., USNR | 45714 |
| FT-117 | CAPT George W. Stivers, USMC | PVT Robert P. Gruebel, USMCR<br>SGT Robert F. Gallivan, USMC | 73209 |
| FT-81 | 2D LIEUT Forrest J. Gerber, USMCR | PFC William E. Lightfoot, USMCR | 46325 |

**NOTE:** The rank of *Captain in the US Marine Corp* is equal to the rank of *Lieutenant in the US Navy*. The technical language for this is *03 Grade Officers.*

**Background Information:** A TBF Navy Avenger can travel a maximum distance to its target of 260 miles and back with two 500 pound bombs or one torpedo. Thus the aircraft has a maximum distance of 520 miles. This distance could be expanded another 20 to 60 miles without the payload of explosives.

## *FACT: Judge Graham Stikelether (Main Figure in Story)*

Former Indian River County Judge Graham Stikelether Jr., at 78 died quietly in 2009. He was born in Kansas and grew up in Tallahassee. His father was a bricklayer who wanted him to be a doctor. Instead, Stikelether graduated from the University of Florida Law School. In 1961, he came to Vero Beach at the request of Sherman Smith Jr., an attorney and former state representative.

Early in his career, Stikelether was a judicial assistant to Smith when Smith was on the 4th District Court of Appeal. Stikelether also was a legislative assistant to former Sen. Merrill Barber of Vero Beach.

Stikelether then started his private law practice in 1965 and served as the county's only judge from 1972 to 1984. After serving as judge, he went back into private practice until 2002.

***In his private life he had a passion for history***. He helped found the Indian River Historical Society and was an admirer of Civil War Gen. Robert E. Lee, said his daughter, Debbie Kanehl of Vero Beach. Because of this ***passion for history***, and while hunting game, he accidentally stumbled onto FT-36...*in **1963***. The fact that he spent a good portion of his life trying *to find the rest of Flight 19* gives validation to what we are presenting.

## The Story:

On the fateful evening of December 5, 1945 when Flight 19 went missing, four radar stations picked up *supposedly* several blips headed towards the Okefenokee Swamp in Georgia. One radar station was located at the Brunswick Airport in Georgia. The second was a National Guard Air Station in Georgia. **The third was the Jacksonville airport in Florida.** The last was a naval vessel whose name we could not find presently. Because of this, some believe that Flight 19 is somewhere in the eastern side of Okefenokee Swamp. No one has ever followed up on this information...*at least to our knowledge*. However, part of this swamp is in a government restricted area. Part of the swamp has been used for jungle warfare training as well as Army Ranger training,...but for the Vietnam War,...*and not World War II* There is no other explanation given as to why this is a restricted area.

The distance from Fort Lauderdale up the Florida Coast to Georgia is approximately 585 Miles. Taking into consideration its practice bombing run, which put it across from West Palm Beach and with an empty payload of sacks of sand (which do not weight as much as a full payload of bombs, thus could have given them a greater fuel capacity) it may have had sufficient fuel to travel that distance.

With their compasses malfunctioning, the airmen became disorientated and confused by a sudden storm that quickly rolled in. We believe they may have headed *in their confused thinking* north by northwest assuming somehow they were in the Florida Keys. There is some conjecture whether their compasses may have started working again once they were out of the influence of the Triangle. The three radar stations mentioned above said that their last radio communication mentioned being over land.

However, we believe far earlier one or two aircraft broke away from the formation (for reasons unknown) and did separate controlled crash-landings on the Florida coast. This would account for Captain J.D. Morrison of Eastern Airlines seeing the red flares rising into the night sky from an island while flying 10 miles south of Melbourne, Florida.

**FACT:** *During the search, Captain J.D. Morrison, an Eastern Airlines pilot, saw red flares rising into the night sky while flying approximately 10 miles south of <u>Melbourne, Fla.</u>* The airline pilot <u>***thought***</u> that they were coming from a small island near the town of Sebastian, of which there were hundreds of these small islands in that area. Captain Morrison agreed to lead a team to the site, where he witnessed the Navy's procedure for a careful search of the area. A careful search consisted of a single helicopter that flew three passes over one island that was surrounded by marshy terrain.

No ground units were involved in the effort, and the thick marshes could have easily prevented searchers from locating any of the airmen who might have crashed in that area–especially if they were unconscious or unable to signal for help. The board criticized the individuals who had been in charge of search operations–<u>*that ultimately resulted in the demotion of several high-ranking officers, including one admiral.*</u>

## The Story goes on:

The rest of Flight 19 then (probably unknowingly) proceeded to their final destination: the Okefenokee Swamp in Georgia. At his time, local news papers later reported, several "Ham Operators" were also tracking Flight 19 transmissions that concurred with the radar stations.

The story of Flight 19 in 1945 doesn't end there. A person claiming to be *Navigator George Paonessa* (at 10:15 a.m.) on December 26, sent a telegram to *George's* family saying he was not dead...and signing it—*Georgie*. His family stated that Georgie was his nickname and did not know if someone was playing a cruel joke on them...*or if it was real*. The telegram *was traced back to the Navy Station in Jacksonville Florida*, however; no other information was available in finding who exactly sent it. (A copy of the actual telegram appears below.)

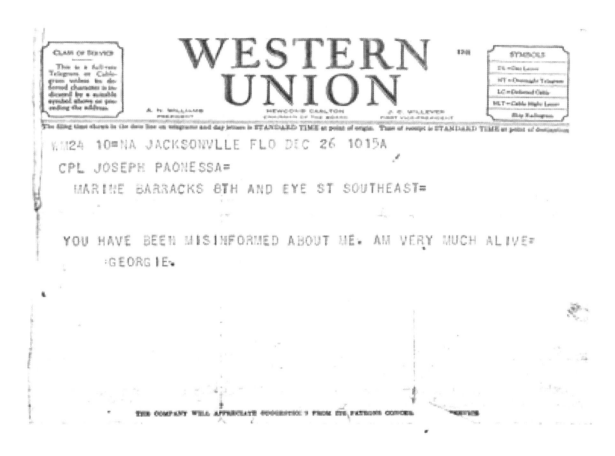

Probably the reason why the family first thought this might have been a cruel joke being played on them was the rank stated. <u>Sergeant</u> (SGT) George R. Paonessa USMC was listed as a <u>Corporal</u> on the telegram. However, we do not know if George R. Paonessa did survive the crash—along being reduced in rank as with the search team "*that ultimately resulted in the demotion of several high-ranking officers, including one admiral.*"

Regardless, the tale of Flight 19 was thought to have ended at this point with nothing else found in the search.

## 1963 - The Continuation of the Flight 19 Story:

Enter Graham Stikelether. While hunting in the nearby swamps in 1963, Stikelether discovered a downed airplane with Naval markings and with two bodies inside. The wreckage was old and thought to be around World War II. It was later identified as a TBM.

**Question:** *Was this TBM aircraft then either FT-81 (which flew with only two crewmen instead of three) or FT-36 where George R. Paonessa might have alone survived a crash landing or some other TBM?*

Upon arriving at home, Graham immediately informed the authorities. When they arrived, he led them back to the downed aircraft. The wreck was then cleared and the two bodies removed. The Navy personnel didn't say much, but one did have a slip of the tongue...*and mentioned that the aircraft must be related to the Flight 19 incident.*

With the aircraft now in the hands of the Navy and with his passion for history, Graham Stikelether contacted the Navy and requested to know what they had found. However, the Navy denied that this incident ever occurred. So he tried other means to satisfied his growing curiosity by attempting to go though other agencies and individuals.

When Graham Stikelether ran out of official channels, he finally contacted a good friend who worked in the Pentagon and asked if he could look into the matter. However, the answer to Stikelether's inquiry was not what he expected; the friend called back and told him to seriously "*drop the case.*" This only confirmed that something was going on, something the Navy didn't want the public to know.

## The Finding of the .50 Caliber Machinegun:

These drawbacks, however, did not stop Graham Stikelether. He persisted in routing out information where he could and made a vast number of treks to the area of the downed aircraft as best he could remember in the late 1960s and early 1970s. He did discover various parts on an aircraft. But his best reward was stumbling across a .50 Caliber wing mounted machinegun with a bent barrel. Obviously the force required to bend the barrel is consistent with wing separation during a crash-landing.

The machine gun had a serial number of either 1055952 or 7055952, unclear as the number was faded on the barrel. The machinegun is now a part of a private collection. But it didn't end here.

**NOTE:** We believe the Serial Number is "1055952", which is correct for that period since finding other .50 Caliber machineguns marked for World War II as 1XXXXXX.

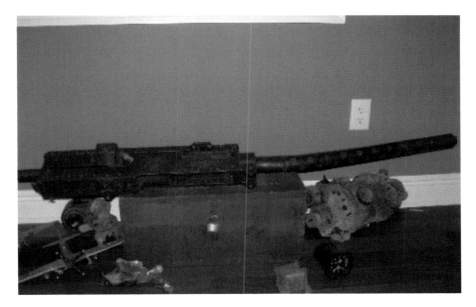

**Photo Credited: Jon Myhre, Author of the book *"The Discovery of Flight-19"***
**(The serial number was later confirmed as belonging to FT-36.)**

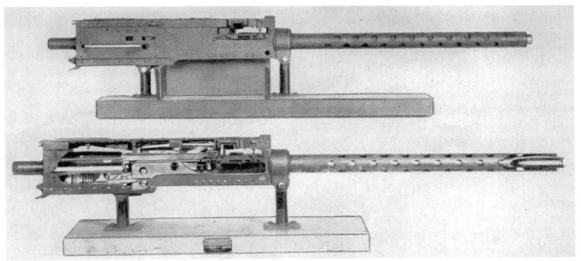

Top: Aircraft Machine Gun, Cal. .50, M2, Fixed. Bottom: Aircraft Machine Gun, Cal. .50, M2, Fixed (Sectionalized).

In his book "*The Discovery of Flight-19*", Jon Myhre details the wreckage of a naval World War II era aircraft found in the wooded swamps near the everglades in Fellsmere, Florida sometime in the 1960's or 70's.

*Another authentic* **.50 Caliber Machinegun from World War II.**
<u>*Almost*</u> **the same type of machinegun mount that was found by Stikelether.**

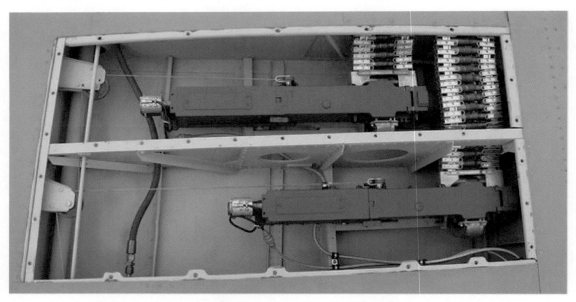

**Same Type of .50 Caliber Machinegun mounted in an aircraft wing.**

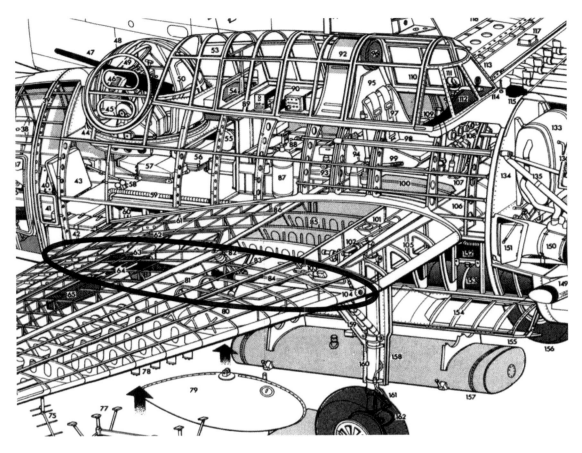

**Cutaway where .50 Caliber was mounted in the wing of a TBM Avenger**

## *The Teaming up of Graham Stikelether and Tom Myhre*

A reporter by the name of Tom Myhre investigated the case in the 1980s, and in 1989, published a story based on Myhre's research in the science magazine *Omni*. After Stikelether seeing and reading the article, contacted Tom Myhre in the hope of learning something more about Flight 19. Myhre, on the other hand, found out through Stikelether about the wreckage found in 1963 and the two bodies aboard.

The two men joined forces in an attempt to solve the mystery and the identities of the bodies found near Sebastian, Florida. However, since Stikelether was unable to provide the exact date of the wreckage recovery, the Navy refused to give any further information.

**NOTE:** *In 2013, Tom Myhre filed a Freedom of Information Act request only to be rejected with an explanation that the names of the airmen found in Sebastian were **redacted** from the Navy recovery files.*

Nevertheless, Myhre realized that even though the squadron was under the command of the Navy, it included several Marine Corp crewmembers. From this he was able to continued his investigation, building upon his theory, and finally concluding that the aircraft found in Florida was the one flown by Marine Capt. Edward Powers, which included Sgt. Howell Thompson, the gunner, and Sgt. George Paonessa, the radioman/navigator/gunner.

Thus it appears that at least one aircraft of Flight 19 went down in Florida while the remaining aircraft went down near or in the Okefenokee Swamp *and not abducted by UFOs* as Mr. Spielberg indicated in his movie. (See maps of suspected crash sites below.) Some of this information was disclosed on the Television show Histories Greatest Mysteries.

**NOTE:** Several different people familiar with that area of the Okefenokee Swamp in Georgia have reported seeing a partially submerged World War II aircraft matching the description of a TBM Avenger.

**NOTE:** *The family did believe the telegram was authentic, as only close family members knew his nickname...**Georgie**.*

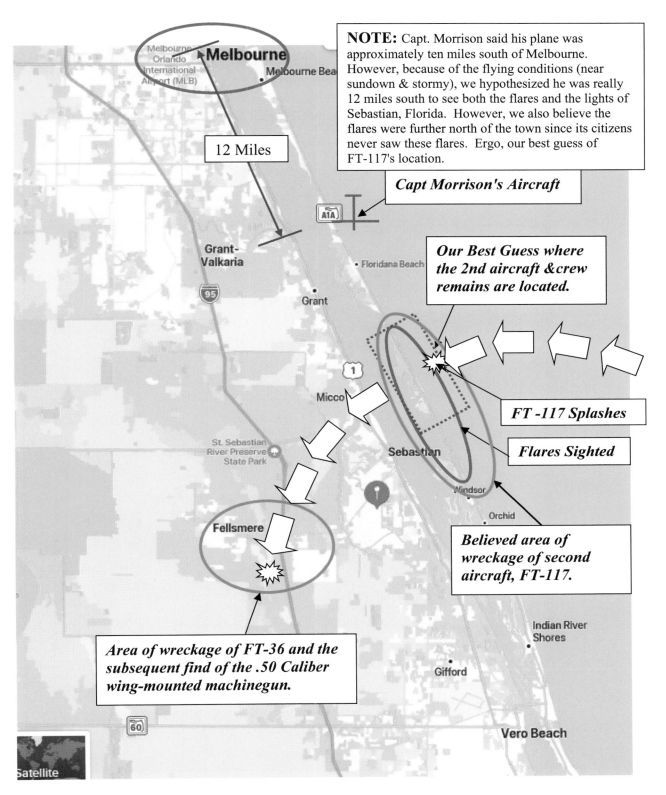

*The First irony of this story was where FT-36 & FT-117 turned into Florida, the naval station from which the Mariner Sea plane took from and then exploded 20 minutes later*

*was only 25 miles north of them. The second and bigger irony was that LT, Taylor's aircraft passed near Jacksonville's airport...and could have landed safely.*

Using a *modern* magnetic declination map of Florida and a protractor, we believed that Flight 19 broke into two flight elements. We then charted both flights' courses and came up with *some astounding results*. ***From both points A & B, each flight turned to a heading of 213 degrees magnetic. This cannot be a coincidence!***

**NOTE:** There is a difference between *TRUE NORTH* and *MAGNETIC NORTH*. *True North* is a geographical location that never changes. However, *Magnetic North* is based on the magnetic poles that change in both position and polarity. Therefore, the course of

213 degrees *magnetic* is an approximation since we used a 2021 magnetic declination map as opposed to the actual magnetic declination map of 1945. Their actual course heading could have ranged between *207 – 220 degrees magnetic* in 1945. *Our objective and intent was presenting that* **both elements of Flight 19 flew parallel to each other on the absolute same heading.**

The end result, Taylor's flight of three aircraft came to crash in a Georgia Swamp while FT-36 and we believe TF-117 crashed around Sebastian Florida.

**Question:** *Did Taylor and other pilots suspect they were off the eastern coast of Florida or perhaps in the Gulf of Mexico?*

**Speculation & Our Assumptions: From the evidence given, it is our absolute belief that Flight 19 was flying parallel to the coast of Florida, pointed at Georgia.** Also based on our combined military training and experiences, we propose that **TWO** aircraft from Flight 19 went down around Sebastian, Florida. These two were FT-36 (Sergeant George R. Paonessa's TBM...***and FT-117, flown by Capt George Stivers, PVT Robert Gruebel and SGT Robert Gallivan.*** The reason being for this,...a US Marine will <u>**NEVER**</u> leave a fellow Marine behind.

It is our contention that FT-117 either developed some sort of engine problem where the aircraft couldn't maintain speed and/or altitude. Captain Powers would have then volunteered his aircraft, FT-36 (to the approval of his crew) and stayed with the lumbering FT-117...or LT Taylor deliberately broke the flight into two sections hoping someone would get back.

There is also a third possible reason; Captain Powers may have disagreed with LT Taylor leadership due to Taylor's inability in making decisions, his disorientation and confusion. Being of the same rank, Captain Powers then assumed command of the Marine element. In any event, Flight 19 split into two flights.

Flight Leader LT. Taylor took his section and flew north by northwest proceeding to Point B with FT-28, FT-3 and FT-81 *above the clouds* while FT-36 stayed behind with FT-117 turned to a heading of 213 degrees magnetic at slower speed and ***below the clouds.*** FT-36 probably had the instructions to crash-land into the sea with FT-117 if it went down.  This was to increase their chances of survival.  Since they flew below the clouds, the storm would have slowed and buffeted the aircraft, making flying more difficult for the ailing FT-117...*and eating up precious fuel faster*.

Although they were flying in a storm and near nighttime, FT-36 and FT-117 were closer to the ground and could visually see of any lights that might appear on the horizon.  Because of this, they must have spotted the city lights of Sebastian, Florida...*and realized they were near land*.  They also would have tried to contact Flight Leader Taylor, but Taylor's aircraft were far out of range as well as the storm causing added radio interference.

Within 20 minutes of this above event, Taylor made his fateful turn to the heading of 213 degrees magnetic, immediately Ft. Lauderdale lost all radio contact with them at 17:50 hours (or 5:50 p.m.).

As the FT-36 and FT-117 closed on Sebastian in the distance, they would have vectored in slightly nearer to determine where to land and the approach to make.  Regardless, they needed to circle around from the north.  As they went into and made their turn, FT-117 had a sudden loss of power and was forced to ditch into the sea near an island just north of Sebastian.

We believe that at least two of the crew survived the crash and swam to an island.  The (injured?) airmen settled in as best they could until they heard as another aircraft above and saw its running lights.

This is where **Captain J.D. Morrison of Eastern Airlines comes in.**

**The survivors fired up their red distress flares, which were spotted by Captain Morrison.**  Unfortunately, the half-hearted rescue attempt never found them...and the survivors died of trauma, and lack of food and water.

As for FT-36, it was probably attempting to circle around to the downed FT-117, but somehow dipped in altitude and wound up diving into the swamp...with Sgt. George Paonessa the only survivor of the crash.  He then probably stumbled his way to Route 1 (*since Route 1 only existed back in 1945 and I-95 wasn't built until the 1960s*) and was picked up,...eventually getting back to Ft. Lauderdale (off of Route 1) Naval Air Station to be quickly transferred to Jacksonville, Florida. He was then probably hospitalized for a short time while he gave his report of the entire affair from his Point-of-View. For some reason he was reduced in rank to Corporal...*to hastily send a telegram to warn his family...that he may not be coming home...at all!*

**Question:** *What circumstances would cause the Navy never to release a Marine who survived this air disaster? If the Navy truly did so, it would have to be a major National Security issue. Was a UFO involved in the disaster after all?*

The telegram is interesting because it speaks of the secrecy that included George Paonessa. He knew he was about to be declared legally dead...and perhaps never see his family again.

Needles to say, Taylor and the rest of Flight 19 died a few hours later, unknowingly passing justly slightly to the north of the safety of Jacksonville airport...to crash into a Georgia swamp. Around the same time, Martin Mariner PBM-5 (No. 59225) blew up trying to find and rescue them. The PBM-5 was nicknamed the "*The Flying Gas Tank*" by its crews, and notorious for fires and explosions.

**Question:** *Why didn't FT-36 land on Route 1? Didn't they see it? Was Route 1 illuminated enough? Was Route 1 wet with rain mistaken for a river or gulley filled with water?*

**Question:** *Why didn't one of the radar stations that picked up Taylor's flight scramble some fighters to intercept the TBMs? Even though World War II was officially over, they didn't know if that was truly Flight 19 or were possible hostile aircraft looking for one last fight, so why not?*

**Question: *Why did the US Navy deny ever finding FT-36?*** *There is no doubt that this aircraft was found:*

1. There was George Paonessa's telegram to his family—FT-36's only survivor. (Remember the aircraft only had two bodies in it.)

2. There was Captain J.D. Morrison of Eastern Airlines spotting the red flares of the second downed aircraft over the island about 6–12 miles away from where FT-36 crashed into the inland swamp. (It wouldn't make any sense if George Paonessa ran the distance and left the mainland to swim to an island to fire off flares for an airliner he didn't see or knew was coming, would it?  Ergo, there had to be a second aircraft of Flight 19 involved, which crashed near or around an island away from the mainland.)  There is also no other reason for FT-36 to have left the flight formation unless it was ordered or volunteered to do by command decision.

3. There was the Navy personnel who had a slip of the tongue...who mentioned that the aircraft must be related to the Flight 19 incident.

4. There was Graham Stikelether's pentagon friend who told him to, **"...drop the case."**

5. There was the recovered machinegun in its mount by Graham Stikelether.

6. There was the 2013 FOIA (Freedom of Information Act) Request that did produced a document with the redacted names of two bodies the Navy recovered from the site...the report **apparently filed in 1963**.

7. There were the radar sightings of aircraft approaching Georgia from the south coming in from the Atlantic Ocean.

8. There was the US Navy slapping a "Top Secret" classification on the air disaster, which it should not have done so.  What did Flight 19 really encounter?

9. There was the spotting of a partially submerged aircraft fitting the description of a TBM in the Georgia swamp by several different individuals over the years.

**Question:** *What was the true reason behind the US Navy's decision to demote several high-ranking officers, including one admiral?*

*Do you see all the logical connections that tell a quite different story about Flight 19? With all due respect, this tragedy is filled with a load of errors that should be now made public!*

## *Fact: Getting Serious about Sirius - a Misconception*

Spectral Classification Letters are part of a star system's footprint, which identify the star's mass, luminosity and temperature. The hottest and brightest stars fall into the A spectral class. Each of these classifications are then divided into sub classes and are given a range from 0 to 10. The 0 class is the hottest, brightest and has the most mass of these sub classes.

M class star systems are the coolest star systems in the universe and burn the slowest. Our sun, however, is a G 2 class star. G class stars have a life span of 10 to 11 billion years. Our sun is about half way through its life span or about 5 billion years old. Stars that fall into the A through F 4 classes have only a stable burning period of 4 billion years before becoming a red giant. Based on our own earth's history, the star systems that fall between A through F 4 classes, should they have planets, these planets would not be able to develop intelligent life beyond the Paleozoic Era, which does not fall past the development of primitive reptiles. Therefore, these star systems must be eliminated as an intelligent life source.

However, some ancient alien researchers have chosen to ignore these basic facts of astronomy, geology and general rules of evolution. In a recent airing on an "Ancient Alien" documentary, some of these researchers claimed that the humanoids found at the Roswell Crash were (supposedly) from the Sirius star system. They support this claims based on our ancient ancestors holding the Sirius star system in high esteem.

Sirius is one of the brightest stars in the universe. Thus, its brightness would make it appear more important to our primitive ancestors. However, Sirius is a class A 1 star and is younger than our sun. Therefore, it is apparent that if any aliens visiting our ancestors did not come from Sirius, they would not have originated there. Since it would take 5 billion years for a civilization to develop there (based on our own as a model), a class A stars like Sirius would only have a life span of 4 billion years. If ancient aliens came from Sirius, it would have been a waypoint (and possibly a commercial or military outpost) from their own home world.

In the case of Betty and Barney Hill's alien abduction as mentioned previously in this manuscript, the star system of Zeta 1 Reticuli was singled out as the home star system of the aliens perpetrating their abduction. The Zeta 1 Reticuli star system is a class G 2 star and about 5 million years older than our own. They also ignored the fact that the aliens who abducted Betty and Barney Hill matched the description of the aliens found at the Roswell Crash site as described by witnesses to that event. The aliens from Zeta 1 Reticuli have been mentioned in many other reports and many other places. And they are the ones thought to have an agreement with the United States government.

Below you will find a list of some of our star system neighbors along with their spectral rating. It should also be noted that the A class star systems rotate faster than our sun, which indicates possibly they do not have any planets around them else the gravitational pull of these planets would slow the rotation of that star (depending on star mass). There is always the chance of a temporary outpost (for whatever purpose), starfaring would require it.

| Name of Stars | Spectrum Class | Distance in Lt Years | Name of Stars | Spectrum Class | Distance in Lt Years |
|---|---|---|---|---|---|
| Achernar | B 5 | 118 | Gliese 796 | G 8 | 47 |
| Aldebaran | K 5 | 68 | Groombride 34 A | M 1 | 11.6 |
| Alpha Mensae | G 5 | 28.3 | Groombride 34 B | M 6 | 11.6 |
| Altair | A 7 | 16.5 | Hadar | B 1 | 490 |
| Antares | M 1 | 520 | Herculis 14 | K 1 | 50 |
| Arcturus | K 2 | 36 | Ind e | k 5 | 11.2 |
| AZUran Majoria | G 0 | 44 | Kappa Forasci | G 1 | 42 |
| Barnard's Star | M 5 | 18 | Kapteyn's Star | M 0 | 40 |
| Beta Canum Venaticorum | G 0 | 29.9 | Kruger 60 | M 3 | 40 |
| Beta Comae Berenices | G 0 | 27.2 | Lacaille 8760 | M 0 | 12.5 |
| Betelgeuse | M 2 | 520 | Lacaille 9352 | M 2 | 11.7 |
| Caneri 65 | G 3 | 44 | Lalande 21185 | M 2 | 25 |
| Canopus | F 0 | 98 | Lalande 25372 | M 2 | 49 |
| Capella | G 1 | 45 | Leo Minoris 20 | G 4 | 47 |
| Centauri A | G 2 | 13 | Lndi E | K 5 | 35 |
| Centauri B | B 1 | 13 | Luyten 726.8 | M 5 | 27 |
| Ceti | G 8 | 36 | Luyten 789.6 | M 5 | 34 |
| Chi Orionis | G 0 | 32 | Mu Arae | G 5 | 37 |
| Crucis A | B 1 |  | NU 2 Lupi | G 2 | 50 |
| Crucis B | B 0 |  | NU Phoenicia | F 8 | 45 |
| Cygni 61 | K 5 | 34 | Pai Aurigno | G 0 | 49 |
| Deneb | A 2 | 1600 | Phi 2 Ceti | F 8 | 51 |
| Eri e | K 2 | 10.7 | Pi Ursa Majoris | G 0 | 51 |
| Eridani 39 | G 1 | 42 | Piscium 107 | K 1 | 24.3 |
| Eridani 40 | K 0 | 160 | Piscium 109 | G 4 | 53 |
| Eridani 82 | G 5 | 20.2 | Piscium 54 | K 0 | 34 |
| Eridani E | K 2 | 33 | Pollux | K 0 | 35 |
| Formalhaut | A 3 | 22.6 | Procyon | F 5 | 35 |
| Gliese 290 | G 8 | 47 | Rigel | B 8 | 900 |
| Gliese 302 | G 8 | 41 | Ross 128 | M 5 | 34 |
| Gliese 309 | K 0 | 41 | Ross 154 | M 4 | 29 |
| Gliese 364 | G 0 | 45 | Ross 248 | M 5 | 32 |
| Gliese 390 A | G 6 | 45 | Ross 614 | M 4 | 40 |
| Gliese 541.1 | G 8 | 53 | Ross 780 | M 5 | 48 |
| Gliese 59 | G 8 | 53 | Sirius | A 1 | 27 |
| Gliese 59.2 | G 2 | 48 | Spica | B 1 | 330 |
| Gliese 641 | G 8 | 52 | Tau Ceti | G 8 | 11.8 |
| Gliese 651 | G 8 | 53 | Teuri 39 | G 1 | 47 |
| Gliese 668 | G 9 | 40 | Vega | A 0 | 26.5 |
| Gliese 67 | G 2 | 38 | Virginis 61 | G 6 | 27.4 |
| Gliese 75 | K 0 | 28.6 | Wolf 359 | M 6 | 23 |
| Gliese 86 | K 0 | 37 | Wolf 424 | M 5 | 44 |
| Gliese 96 | G 5 | 46 | Zeta 1 Reticuli | G 2 | 37 |
| Gliese 97.2 | K 0 | 52 | Zeta 2 Reticuli | G 2 | 37 |
| Gliese 722 | G 4 | 49 | Zeta Doradus | F 8 | 44 |
| Gliese 788 | G 5 | 49 | Zeta Tucanae | G 2 | 23.3 |

# Chapter X: Notable Movies and Television Specials Recommended for Viewing

**In aiding us in our research while comprising this book, we chose a number of movies and television specials to review, which offered the most detail information that was not only exemplary, but left a trail we could expand upon. We selected only material that held mostly creditable accounts or in depth interviews with witnesses or abductees. These movies in our opinion contain some of the most compelling photographs and film footage of genuine UFOs. Many of these photographs and film footage were taken in broad daylight as oppose to just mere lights in a night sky.**

## *Fact: UFOs Are Real*

In 1990, a documentary was released called "UFOs ARE Real". This movie started off by showing never before seen photographs and 8mm camera footage of Unidentified Flying Objects. It spoke about the Kenneth Arnold sighting along with other sightings that were made around the world. Some of these still photos were later deemed hoaxes along with several movie clips. In other cases some of the still photographs were blurred, out of focus or blackened which made positive identification skeptical. However, several of those still photographs and movie clips were methodically analyzed and could not be explained other than as a being of genuine UFOs.

One set of three still photographs snapped by a Polaroid camera were particularly interesting. This was the Rex Heflin photographs taken in 1965. The UFO was snapped out the window of a car and was near a windmill so it had distance verification. These three pictures were taken with a Polaroid camera, which made it almost impossible to touch up or fake. It was so impressive, a paranormal television show named "Fact or Faked" tried to dismiss them, but couldn't.

However, it was the "UFO Hunters" who proved the absolute validity of these photographs. Not only did they demonstrate it was possible to take the stills in the time allotted by an expert using the same model Polaroid camera, but through state-of-the-art computer imaging and computation proved that Rex Heflin's own calculations were absolutely correct. There is now no doubt that the object was 20 feet in diameter at an altitude of 150 feet approximately 1/8th of a mile away. This cannot be disputed!

As to some of the other film footage shown in "UFOs ARE Real," there was no question that these crafts were genuine Unidentified Flying Objects. This documentary also interviewed Betty Hill, Marjorie Fish and Travis Walton about their alien abduction experiences. The results of these interviews and their stories were believable, although some of Travis Walton's interview was edited. However, their research on Billy Meyers and George Adamiski we considered tainted as both of these claimers were later proven to be hoaxers. This particular documentary is still worth viewing because some of the photographs and movie footage is quite impressive as is the interviews of the true abductees Betty Hill and Travis Walton.

***Fact: Documentary Television Special: UFO Cover-Up Live***

In mid October of 1987 a documentary television special was aired on channel 11 (WPIX) entitled "UFO Cover-up Live". This show had Mike Farrell as host and a multitude of guest including renowned Nuclear Physicist Doctor Stanton Friedman. It also had two individuals from the intelligence community of Area 51 which were code named *Condor* and *Falcon*. Their features were darkened and voices were disguised so that their identity would remain secret until a congressional investigation could be started which this show was calling for.

The producers of this show purportedly verified the credentials of both *Condor* and *Falcon* to insure that they were both government intelligence agents. These two stated that the United States government covered up any creditable UFO cases and these cases were never made available to the researchers of Project Blue Book under the guise of National Security. They stated that the United States government had proof of alien beings both alive and deceased, which were retrieved from the Roswell Crash of 1947 and that they called these creatures EBEs (Extraterrestrial Biological Entities). They further stated that two of the aliens that were alive became part of an exchange program. *Condor* and *Falcon* went on to say that Majestic 12 or MJ - 12 was a real organization, which made policy and procedures on how to deal with Extraterrestrial Visitations. These agents stated that the United States government had made an agreement with these aliens that the United States government would not disclose the alien's presence on earth if these aliens did not interfere with our society.

*Condor* and *Falcon* said that these aliens were from the star system located in our southern hemisphere known to us as Zeta 1 Reticuli and they had been visiting this planet for centuries. These aliens were described as being 4 to 5 feet in height with large heads and almond shaped eyes. The aliens could live anywhere from 350 to 400 years and had an IQ (Intelligence Quotation) of at least 200, which is over a 100% higher than our normal human IQ. *Condor* and *Falcon* stated that these aliens were vegetarians and then described the alien's internal organs. Their heart and lungs were combined together as one organ. Another of their organ's turned solid matter that they consumed into liquid. They did have males and females to reproduce. Their planet was closer to the sun than ours and thus they developed an extra film over their retina to protect their eyes from the sun's rays. There does appear to be some difference in height description of these aliens, however, even among our species this is not uncommon and must be disqualified as a method of discerning the truth.

**Speculation:** This television special only aired once in November of 1987 and was supposed to air a 2nd time the following week at the same time it originally aired with the results of viewers call in for a congressional investigation. However, it never did due to being quelled by a government agency, which was rumored to have threatened the producers of the show and the network with removal of their FCC (Federal Communications Commission) license. Why would the federal government threaten the producers and network unless *Condor* and *Falcon* were telling the truth about alien visitations. It is curious that there have been a limited amount of alien abductions that are not in conflict with the Heisenberg Uncertainty Principle Theory since 1976.

**Question:** *Could the alleged exchanged program have something to do with this?*

**Question:** *Where are Condor and Falcon today?*

**Question:** *Were these 'whistle-blowers' identified and permanently silenced for their indiscretions?*

It is curious that a copy of this show cannot be purchased on the internet and that the only way that an individual can view this show is on YOUTUBE (in part) or at the following website: http://video.anomalies.net/video/123/UFO+Coverup%3F+Live!

**Fact: Alien Secrets - Area 51**

In 1995, a television special was released entitled "Alien Secrets – Area 51". This location is also known as *Dreamland* and is located eighty miles from Las Vegas. The main concept of this documentary covers the rumor that reverse engineering of alien space craft is taking place at this remote facility in an area known as Sierra 4 or S-4.

Interviewing Robert Cechsler NASA (National Air and Space Administration) mission specialist admitted that intelligence sources from Area 51 stated that the United States government does have a flying disc at that location. This documentary also interviewed Lieutenant Colonel Wendel Stevens (retired) from the Foreign Technology Division of Wright Patterson Air Force Base in Dayton, Ohio and a former member of Project Blue Book who admitted that the United States government is in contact with aliens and that the CIA (Central Intelligence Agency) told the panel of Project Blue Book what cover story to use with each genuine UFO report in an effort to mislead the public.

The reporters of this documentary also interviewed a military radar specialist who observed a test of at least 10 saucers piloted by aliens in the 1980's at Area 51. Keeping her identity a secret, she stated that afterwards she was given an injection and suggestion not to speak of the incident. There was also a 71 year old engineer that came forward (hidden in shadow) and explained that he worked in Area 51 on a simulated test training disc from 1965 to 1979 for a satellite government of the United States government.

He stated that there were 4 alien beings helping with the project and that one of them was named *Jay-rod*. This gray alien being supervised the construction of the training disc and would speak with them telepathically.

Area 51 in Nevada was commissioned by former President Truman in the 1950's. The United States government has denied the existence of Area 51 until former President Clinton sign a bill into law that admitted to the existence of this base. During this time period several workers at this location were dying from toxic waste and in this law the government refuses to acknowledge what caused their deaths. It further states that anyone working for the government in this area has all rights removed from them if they disclose what they know about Area 51 and without trial can be sent to Leavenworth Federal Penitentiary for 20 years. This documentary also covered the Bob Lazar story previously mentioned in this book. Also in this special there is an artist conception of the disc shaped craft as described by witnesses.

Reporters of this documentary investigated organizations that are working on government projects such as E G & G ( Edgerton, Gemeshausen and Grier Inc.) Special Projects and SAIC (Sign Applications International Corporation). These companies appear to be working on an antigravity electromagnetic propulsion system, which can bend space and time. These same reporters then drove down Extraterrestrial Highway to Groom Lake road with the intension of displaying the protection that surrounds Area 51 such as the motion sensors, cameras and physical patrols. They gave examples of signs that are posted in this area which read as follows:

**Restricted Area**
**No Trespassing Beyond this Point**
**Photography is prohibited without authorization of Base Commander**
**Punishment $5,000 fine & 1 year imprisonment**
**Authority Internal Security Act 50**
**Authorized use of Deadly Force is Permitted**

This documentary also mentioned Black Projects and elaborated on "Cosmic Journey" a project that was to be presented to the public, which covered the crash at Roswell and showed the saucer and dead alien bodies that were recovered explaining where these beings were from and about their lives. This documentary is an eye opener and should be viewed by anyone interested in Unidentified Flying Objects and Alien Technology.

**NOTE:** *The project "Cosmic Journey" was cancelled by higher officials in the government.*

### *Fact: UFO Files: Alien Engineering*

The History Channel has put out a DVD from their UFO Files entitled "Alien Engineering." This film is a collaboration of scientist speculating on how the alien space craft operate. They start the film saying that they recovered a alien space craft that had crashed in the desert and brought it to a hanger for analysis. (*Sounds familiar?*) They start off discussing the different shapes of these space craft that have been seen over the past 70 years. They debate these different shapes against aero dynamic designs and why one shape is preferred over another. They state that the flying wing or boomerang design is preferred which was the same design craft that was reported by Kenneth Arnold in 1947 and similar to the craft that crashed at Roswell, New Mexico.

They then went into G forces, which are forces of gravity that act upon pilots on acceleration and deceleration. Humans can only with stand up to 9 G (Gravitation Acceleration Force) forces before they are affected by gravity, which will cause the blood to drain from the brain. If this happens then unconsciousness will then occur that will enviably lead to death. This is due to the lack of oxygen in the brain since blood supplies oxygen to the brain. The scientist reasoned that the way UFOs can perform erratic maneuvers such as 90 degree turns or stopping short, which would normally cause a force of 300 Gs inside the craft, is they have an inertia canceller. In other wards the beings that operate these craft have installed a gravitational field inside these craft that act against the gravity of the planet thus canceling out the gravitational force.

In reviewing how the craft can make these sharp turns instantaneously, it was reasoned that the craft was controlled telekinetically. In other words, the ships computer is linked directly to the alien pilot's brain. The thought process of the pilot determines if the ship goes up, down, left, right, reverse etc. This would seem logical since these craft are able to out maneuver our jets in an instant.

**NOTE**: Curiously Lieutenant Colonel Phillip Corso also believed that this was their method of controlling their craft, which he deduced from the head band found at the Roswell crash.

The propulsion system is believed to be some sort of anti-gravity or possibly diamagnetic field propulsion system. (*Diamagnetic materials are those materials that are freely magnetized when placed in the magnetic field. However, the magnetization is in the direction opposite to that of the magnetic field. Water is such a material.*) All living things on our world contain water. Using a diamagnetic field, these scientists levitated a frog by changing the polarity of the creature allowing the water contained within to lift it upwards.

They then covered the power source, which they believed was anti-matter. In their calculations 1 ounce of anti-matter combined with 1 ounce of matter could produce the same amount of energy equal to 1 million tons of rocket fuel. A company called Pherme Laboratories has been in the process of making anti-matter. It is theorized that this power source could also be used by the aliens as a weapon.

The next area covered was traveling through the vastness of space. Professor of Theoretical Physics Michio Kaku from the University of New York described how traveling through space can be accomplished by the use of worm holes. This is done using gravity or an intense magnetic field, which can distort space and time. This was first discovered by Albert Einstein in 1935, which he called a Rosenbridge. Gravity applied to mass can bend space around that mass causing a worm hole between two points thus folding space around that mass and moving the two points closer through this worm hole even though these points have not physically moved from their original location in the universe. Travel time is thus shortened to hours instead of centuries between star systems.

Although this film did not cover actual UFO cases, it did give a unique view as to how these craft operate. It is interesting how the electromagnetic system seems to control so many functions of these craft. It is also theorized that the electromagnetic field of these craft affect the engines of our vehicles as stated in so many UFO reports. Perhaps these scientists' speculations are more close to reality then the audience is led to believe.

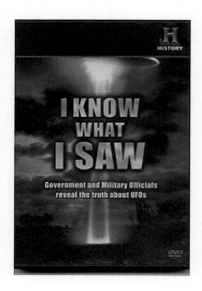

### Fact: I know what I Saw

**"I know what I saw"** is a film produced by James Fox and opens with the Phoenix Lights, a UFO phenomenon that occurred on March 13, 1997, from 20:20 to 21:00 hours. This craft was so huge that some witnesses compared it to Camelback Mountain. Another witness stated that it glided, stopped and hover, then its wings went partially up and took off instantaneously. The FAA (Federal Aviation Administration) and Luke Air Force Base had no explanation for this event. This craft emanated no noise the whole time it was witnessed. Next the film shows photographs of UFOs some of which have never been seen before. These presentations include: Santa Ana California Photo (clear daylight Photo) of 1965; Salida, Colorado Film of August 27, 1995 in which on a clear day this UFO was filmed and one can clearly see lights going around the craft; and the Vancouver Island photo of 1981.

The film then covers the UFO researchers that organized a meeting at the National Press Club that took place on November 12, 2007; which called for the United States government to open all its files on Unidentified Flying Objects. The National Press Club covered UFO sightings from around the world starting from World War II. It mentioned the UFOs that flew over the White House in Washington in 1952 and how radar estimated their speed at over 7,000 miles per hour. In that same year Life Magazine covered the subject of UFOs where some Generals in the United States Military were quoted saying that they were convinced that these UFOs were from another world.

This film also covered Doctor J. Allen Hynek last interview were he admitted that the CIA (Central Intelligence Agency) controlled the outcome of Project Blue Book in that the project researchers were given a set of criteria with which they must use to dismiss most of the UFO sighting that were reported. He also stated that any sighting, which could not be explained away, was not given to the researchers at Project Blue Book. Since Project Blue Book was closed, there have been over 25,000 reports of UFO encounters. An internet survey of 2007 stated that 83% of the people that responded to this survey believed in UFOs.

The movie also covered Nick Pope's investigation of the Rendelsham Forest Sighting near Bentwaters Air Force Base in the United Kingdom. In his investigation he stated that the radiation reading in the area where the triangular craft was seen was 8 times higher than normal and that the craft had made 90 degree turns.

This film also covered the UFO report of a giant craft seen by the crew of Japanese flight 1628 which took place on November 17, 1986. This craft was said to be 4 times the size of the 747 aircraft in which the Japanese crew was flying. The flight controller of the tower, which received the radio transmission from the pilot of the Japanese flight, stated that the CIA came in and confiscated all the material (data) with regards to this report and said that they (the CIA) told everyone that they were never there and that this event never happened.

This film also spoke of the "COMETA Report" that was released by France with regards to UFOs around the world and what governments should prepare for. I viewed this film with a UFO critic who was visually disturbed by the evidence presented in that he was afraid that aliens would take over the earth. This movie is the predecessor to the movie "Out of the Blue" and is a must to see and own for anyone interested in this subject matter.

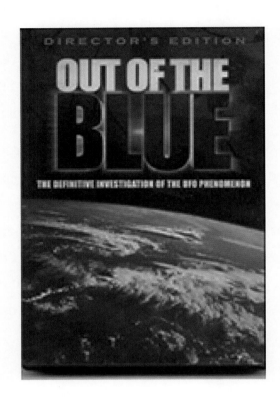

### Fact: Out of the Blue

**"Out of the Blue"** was a documentary that was released in 2009 / 2010, which covered a multitude of subjects on Unidentified Flying Objects. It opens with some incredible photographs and film clips from all over the world. Some of the more fascinating presentations are photos and film footage taken from the following: 1956 California film; 1964 Merlin, Oregon film (a hovering disc clearly visible in broad daylight); 1976 Hasenbol, Switzerland; 4/03/1990 Krasnadur, Russia; 3/31/1991 Mir Space Station film; 9/09/1995 Ecuador film (red object hovering) and 7/14/1996 Vaxjo, Sweden.

This documentary also covered some of the most controversial UFO incidents on record, such as "The Phoenix Lights," which had over 800 sightings reported between 20:15 and 20:45 hours. In this sighting it was stated by witnesses that a gun metal black color V-shaped craft was so huge that it blocked out the stars as it passed over. One witness estimated the size to be around 1400 feet in diameter. Another witness said that the object was so large that the United States military could park forty B- 2 bombers on its wings.

The channel News 15 agency in Arizona checked with Luke Air Force Base and they had no explanation. However, later the government stated that there was some sort of military practice that night and that these lights were flares. In checking the military did drop flares that night, but that was more that 2 hours after the UFO had vanished. Another witness who was making a delivery to the air force base stated that he had observed two smaller orbs earlier over the airbase. He further stated that the military dispatched two fighters that went into after burners in pursuit of these orbs. The flight path of the giant UFO was said to be from Prescott to Tucson Arizona.

Another case that was covered was a UFO attacking a Atlas missile during an experimental launch. The UFO was recorded on film shooting a blue beam and blowing up the warhead. The film also covered in part the Bentwater / Rendlesham Forest Incident. In this case a UFO came over a nuclear airbase in England on several nights. In this incident three security personnel were dispatched and came upon a triangular shaped craft in the forest outside the base. They interviewed Admiral Lord Hill Norton former Chief of Defense Staff for the United Kingdom about this incident who stated that he believed this event really occurred.

Especially interesting was the film from Somerset, England were during a daylight observation a silver disc UFO was hovering above a communication tower on 7/04/1998. This footage is known as the Dickinsen Film and was brought to Lucas Film studios. Here it was analysis by Bill George in special effects photography. He stated that this was not computer graphics as it went in and out of focus with the rest of the film, which was almost impossible to duplicate. He further stated that he believe that this object was really there. In all "Out of the Blue" is not only worth viewing, but owning for anyone interested in UFOs. It covers much more than mentioned here and is quite an eye opener that would give most critics a disturbing reality check.

# Chapter XI: An Analysis And Still More Questions

## *Analysis*

In covering several different segments during our research it became apparent that two of these star systems were continuously being referred to in one manner or another. These star systems are recognized in either by description of the aliens that are supposedly to inhabit that system or by their location in the universe according to the Gliese Star catalogue. These stars which are mentioned in several of the incidents are Zeta 1 and 2 Reticuli.

From early in our history, these beings seemed to have been present on our planet. An example of this can be seen by the skeletal remains found where the skulls were elongated. Our ancient ancestors from both Central and South America and Africa would wrap their newborn children's skulls in order to elongate them to match their celestial visitors (which they presumed were gods) as a sign of honor. These rituals can also be seen in drawings and paintings found in tombs and on methodical structures.

Descriptions of these aliens were also given by the witnesses at the Roswell Crash and by both Betty and Barney Hill and Travis Walton which matched these paintings. The star map that Marjorie Fish created from Betty Hill's hypnosis session gave only one matching stars systems, which were Zeta 1 and 2 Reticuli.

Intelligence agents from Area 51 code named *Condor* and *Falcon* admitted that the aliens found at the Roswell Crash site were from Zeta 1 Reticuli. Atmospheric conditions described by Travis Walton during his abduction aboard the alien craft point towards a heavier carbon dioxide content, which is in keeping with a planet that has larger photosynthesis occurring that indicates a greater concentration of plant life. *Condor* and *Falcon* stated that these aliens from the Roswell crash were vegetarians, thus it can be presumed that they are an agricultural race. This means that they would have a higher content of carbon dioxide in their atmosphere because of the greater amount of plant life. Finally Zeta 1 and 2 Reticuli fall into the G spectral star system and are estimated to be 5 million years older than our sun. This is the correct star spectral that would contain planets and thus has a greater ability to have developed intelligent civilizations and being millions of years older these civilizations on Zeta 1 and 2 Reticuli would most likely have conquered space travel.

It is curious that in the 1940s, these UFO or "*Foo Fighters*" that were most often seen were small in size. However, throughout the years they seem to have graduated to those of huge proportions as in the phoenix lights and in the case of JAL flight 1628. In 2012, the Discovery channel introduced a new show they called "Curiosity." On one of the opening shows they did a speculation on how an alien culture would invade the earth.

They interviewed military and the scientific communities on their best conception for an attack upon this planet. The show stated that these advisors estimated that the preliminary attack would be in the form of artificial asteroids being propelled at great speeds into our oceans near all the major ports and cities to create giant tidal waves. Since 80% of the earths population lives by the water this would most likely wipe out over 60% of the humans on this planet.

The enemy next attack would be in a form of an unknown viral infection, which would be spread by birds or bats. This would most likely eliminate another 15% of the remaining population. What would be left would be scattered and unable to mount a united defense. They estimated that the enemy would only need a few large space craft to perform this invasion. We would be almost defenseless since our space shuttle does not support a weapons system and the only other aircrafts that could possibly retaliate would be the SR 71 Blackbird or the mythical Dark Star "Aurora Project". Our missile system would be too slow to reach them and could easily be detected by their more advanced warnings systems.

It is already known that their craft have an electromagnetic force fields, which would repel most metal, thus our projectile casings would need to be constructed from titanium, which is very expensive, but is not affected by magnetic fields. This whole scenario appears extremely alarming. However, according to the rumored landing at Holloman Air Force Base in White Sands, New Mexico; it appears that these aliens are more interested in creating a trade agreement with our world. This particular information was also mentioned by Betty Hill that aliens from Zeta 1 Reticuli had established a trade route with other planet colonies such as 82 Eridani and Alpha Mensae, both of which are in a G 5 spectral star system class.

"Cosmic Journey" was a production that was supposed to end the UFO controversy. It was designed to introduce the American public to alien beings that were visiting our world. This production was to incorporate the extraterrestrial beings from Zeta 1 Reticuli that had crashed at Roswell, New Mexico showing their physical structure and their space craft. It was to include information about their society, their visitations to our world, their explorations of space and other societies on other worlds. This was to be presented in a manner as to avoid the panic that occurred in 1938 when Orson Wells broadcast the *"War of the Worlds"* from the Mercury Theater.

Then this project was cancelled by the government. Why, no one appears to know. Perhaps the powers that be at that time may have thought the general public still could not handle the truth about alien visitations. However, "Cosmic Journey" is still a better solution to introducing these alien visitations to the general public then having the truth be revealed by an opposing nation and condemning our government in a conspiracy to hide the exchanging of technologies with these aliens and keeping this knowledge from the world.

**"WE, THE PEOPLE"** have a right to know! After all, does not the Constitution of the United States make it absolutely clear that those who serve in the government are supposedly to be servants of the people? And who has the right to deny you (or anyone else for that matter) the candid information needed to make individual decisions for the necessary precautions in safeguarding you and your family's welfare as well as the public good through majority rule? No bureaucratic or politician can possibly know each and every individual situation or truly claim to speak for the people when they collude in secrecy.

So what it all comes down to is this: the violation of our God-given Constitutional rights. And these rights have been paid for by not only our tax dollars to a wasteful, bloated, and indifferent government, but by the blood of this country's finest sons and daughters in fighting wars. We have every right to know the truth! And the Constitution is not just a piece of paper. It is the rule of law…our law by which this country stands by! And we should be become very, very vocal in demanding answers…truthful answers…and in holding these bureaucrats accountable for any crime, any indiscretion, any stupid decision made in the name of "National Security," which seems nothing more to be a catch-phrase of government wastefulness and abuse.

## *More Questions*

**Question:** *In regards to SETI, if these signals would break up before reaching any star system and the scientist knew this then why would they even create this program?*

**Question:** *Are SETI funds actually being used to finance black ops programs? Where is the money going and for what?*

**Question:** *Why haven't the clear and later remains of our ancient ancestors been found?*

**Question:** *What about the pyramids in Egypt and South America, how did people separated by thousands of miles conceive and design virtually the exact same structures?*

**Question:** *Then there is the engraved stones which were carbon dated back to the Mesozoic Era with depictions of both man and dinosaurs on these same stones, how is this possible with 65 million years separating these species?*

**Question:** *What of the rock that was found in Glenrose, Texas that contains footprints of both a dinosaur and a human imbedded in it as the lava cooled, if these two species are separated by 65 million years how do the critics explain this when scientist that examined it could not?*

**Question:** *Finally the skull of the Star Child which shows from DNA testing that the father of this child was of non human origin, how is this possible without alien intervention?*

**Question:** *In the Battle for Los Angeles, there is absolute proof that these were not conventional aircraft since their physical profile appeared as saucer-shaped in the few photographs taken and could not be shot down by accurate anti-aircraft fire. What were they then?*

**Question:** *How do the skeptics explain the more than 20 deathbed confessions by military and government personnel involved with Roswell as actually had happened?*

**Question:** *What about all the people of Roswell who witnessed the saucer, the alien, bodies or handled the debris? Were they all hallucinating?*

**Question:** *If so, why were all these people threatened with death then by the United States Military if they revealed what they observed?*

**Question:** *Why would the government of the United States threaten to kill their own citizens over the observation of dummies or Chimpanzees which is what the government claims these witnesses saw at the Roswell incident?*

**Question:** *With the civilian news reporting agencies, why were they all told that they would lose their Federal Communications Commission license if they continued to broadcast any information about the Roswell Crash of an alien space craft when the government clearly stated it was a balloon?*

**Question:** *How would Betty Hill known about a medical procedure that was not yet invented?*

**Question:** *Why would the Hills descriptions of aliens appear to be identical to the ones described in the Roswell crash (that they didn't know about) unless these beings were from the same planet?*

**Question:** *How would Betty Hill know about this exact star position of Zeta 1 and 2 Retucli from Marjorie Fish's star map reproduction if she had no knowledge of astronomy?*

**Question:** *Why are Zeta 1 and 2 Retucli constantly mentioned by university studies in good possibility of containing extraterrestrial life as well as coming up in UFO contact cases? How do the skeptics explain this?*

**Question:** *If Travis Walton did not experience an alien abduction, how did he know about the difference in the thickness of the air quality matching the Roswell story's claims the gray aliens found it difficult to breathe our atmosphere if the Roswell story did not unfold until several years after his abduction ?*

**Question:** *What did Captain Thomas Mantell's wingmen observe and why were their interviews never released by the military?*

**Question:** *What about the craft seen a few hours later at Lockbourne Air Field in Ohio?*

**Question:** *What happened to the eye witness reports from Lockbourne Air Field in Ohio?*

**Question:** *How does the government explain the UFOs over the Capital in Washington DC in 1952?*

**Question:** *What about Malxstrom Air Force Base in Montana as reported by Lieutenant Colonel Salas, how does the Air Force explain the complete loss of launch control systems for over 30 nuclear weapons when one of these non existing Unidentified Flying Objects commanded the sky over this base?*

**Question:** *About the movie footage taken at Holloman Air Force Base and the first face-to-face encounter with aliens, why (if this was theatrical footage) did they use the actual base commander to greet the aliens?*

**Question:** *Why wouldn't the Air Force give a copy of this footage to Paul Shartle and Robert Emenegger if this was only theatrical footage?*

**Question:** *In the Cash - Landrum incident, how do the skeptics explain how these two women contracted radiation poisoning?*

**Question:** *JAL flight 1628 reported a craft 4 times the size of their 747, what could this have been since no country has any aircraft or airship that large?*

**Question:** *About "UFO Cover – UP Live," did the government close down this show?*

**Question:** *If so was it because it was calling for a congressional investigation or because the two intelligence agents were exposing the truth about Alien visitations?*

**Question:** *Who cancelled the "Cosmic Journey" project and why?*

**Question:** *How do the critics explain the clear daylight footage of a flying disc given from not only the COMETA report from France, but the movies: "I Know What I Saw and Out of the Blue"?*

**Question:** *Why did the US military aircraft drop flares over Phoenix, Arizona after a large UFO was sighted two hours before? Was this a cover up for a real UFO occurrence?*

**Question:** *In another case dubbed the "Somerset Film," which was taken in England, this film is of a silver disc in daylight, which was hovering over a communications tower. The producers of this film wanting to validate the footage took it to a film laboratory for analysis. The film laboratory concluded that the craft which is being observed in the film was actually at that location and not a computer graphic. This footage is so clear that light spinning around the craft can be seen. How do the critics explain this?*

**LAST and FINAL QUESTIONS: If the United States government has an agreement (*as testified by various sources presented within this book*) with Extraterrestrial Aliens, what does the specifics of this deal entail? For what...*or whom* are these aliens trading their technology for? And what is their ultimate purpose? AND IF ALL is TRUE, why was this agreement not disclosed and voted upon by the United States Congress as specified by the Constitution of the United States, which clearly defines such an undisclosed agreement as being both *illegal* and *criminal*? Is it any wonder why the United States government would want to keep the existence of UFOs and alien visitations a very closely guarded secret?**

The skeptics are looking for proof that our world is being visited by Aliens in strange spacecrafts. We have just presented eye witness testimony, government agent's confirmation, films to review and material with incidents, which cannot be explained away. If they really wanted proof then they should do the research and they will find that it is starring them in the face.

The fact of the matter is that they really don't want to believe that there are more intelligent beings in this universe than us and that these beings are visiting this planet. This is possibly due to these skeptics own insecurities and fears. It is either that or these skeptics are being told by the government of the United States to cover up these visitations of alien craft with lies, fallacies and subjugation of the truth. This still doesn't change the fact that all this evidence exist. It is better to keep an open mind and be prepared for any circumstance than to have a closed mind and be overtaken by the unexpected.

We have already presented the facts that we have found in our research. We have drawn our own conclusions from those facts. However, we do not ask you the readers to take our view without reviewing the proof for yourselves. We only ask that you read the material suggested and view the films and television documentaries mentioned, and then draw your own conclusions. You decide if **"*Rumors of UFOs*" and "*Alien Encounters: The Truth*" are Fact or Fiction!**

Printed in Great Britain
by Amazon